Regression Estimators

A Comparative Study

This is a volume in
STATISTICAL MODELING AND DECISION SCIENCE
Gerald J. Lieberman and Ingram Olkin, editors
Stanford University, Stanford, California

Regression Estimators

A Comparative Study

Marvin H. J. Gruber

Department of Mathematics
Rochester Institute of Technology
Rochester, New York

ACADEMIC PRESS, INC.
Harcourt Brace Jovanovich, Publishers

Boston   San Diego   New York
London   Sydney   Tokyo   Toronto

ACADEMIC PRESS, INC.
1250 Sixth Avenue, San Diego, CA 92101

United Kingdom Edition published by
ACADEMIC PRESS LIMITED
24-28 Oval Road, London NW1 7DX

Library of Congress Cataloging-in-Publication Data

Gruber, Marvin H.J.
    Regression estimators : a comparative study / Marvin H.J. Gruber.
    p.   cm. — (Statistical modeling and decision science)
    Includes bibliographical references.
    ISBN 0-12-304752-8 (alk. paper)
    1. Ridge regression (Statistics)   2. Estimation theory.
  I. Title. II. Series.
  QA278.2.G78   1990
  519.5 — dc20   89-29740
                                                             CIP

Printed in the United States of America
90 91 92 93   9 8 7 6 5 4 3 2 1

Part IV   Applications

Contents

Part II   The Estimators

Part III   The Efficiencies of the Estimators

Preface

The study of different mathematical formulations of ridge regression type estimators points to a very curious observation. The estimators can be derived by both Bayesian and non-Bayesian (frequentist) methods. The Bayesian approach assumes that the parameters being estimated have a prior distribution. The derivations using frequentist type arguments also use prior information, but it is in the form of inequality constraints and additional observations. The goal of this book is to present, compare, and contrast the development and the properties of the ridge type estimators that result from these two philosophically different points of view.

The book is divided into four parts. The first part (Chapters I and II) explores the need for alternatives to least square estimators, gives a historical survey of the literature and summarizes basic ideas in Matrix Theory and Statistical Decision Theory used throughout the book. Although the historical survey is far from exhaustive, the books and research papers cited there should provide the interested reader with many entry points to the vast literature on ridge type estimators. The author hopes that the summaries of results from Matrix Theory and Statistical Decision Theory not usually presented in undergraduate courses will make this specialized book readable to a wider audience.

The second part (Chapters III and IV) presents the estimators from both the Bayesian and from the frequentist points of view and explores the mathematical relationships between them.

The third part (Chapters V–VIII) considers the efficiency of the estimators with and without averaging over a prior distribution.

Part IV, the final two chapters IX and X, suggests applications of the methods and results of Chapters III–VII to Kalman Filters and Analysis of Variance, two very important areas of application.

In addition to the theoretical discussion numerical examples are given to illustrate how the theoretical results work in special cases. There are over 100 exercises to help the reader check his or her understanding.

The author hopes that this work will prove valuable to professional statisticians and workers in fields that use statistical methods who would like to know more about the analytical properties of ridge type estimators.

Any work of this kind is not possible without the help and support of many people. Some of the research that led to the writing of this book began in the author's Ph.D. dissertation at the University of Rochester (Gruber, 1979). The author is very grateful to his thesis advisor, Dr. Poduri S.R.S.Rao, for suggesting dissertation work in this area, his help, encouragement, prodding, and patience. A lot of benefit was also derived from later collaboration with him, especially on the joint paper, Gruber and P.S.R.S.Rao (1982).

A number of colleagues at the Rochester Institute of Technology provided help and support in various ways. These include Dr. John Paliouras, Dr. George Georgantas, Prof. Pasquale Saeva, Prof. Richard Orr, Dr. Patricia Clark, Dr. James Halavin, Prof. David Crystal and Dr. Mao Bautista.

The author is grateful for the financial support granted for his research in the form of a Dean's Fellowship in the College of Science during several summers, and for released time from some of his teaching duties to work on the book.

The first draft was prepared by Stephen Beren of Kinko's Copies. The final camera ready version was typed by undergraduate students in the Applied Mathematics, Computational Mathematics, or Applied Statistics programs of the Mathematics Department of Rochester Institute of Technology's College of Science. The students include Richard Allan, Christopher Bean, Elaine Hock, Tanya Lomac and Christopher J. Mietlicki. The quality of this book was greatly enhanced by their painstaking efforts.

The author is also indebted to the secretarial staff of the Department of Mathematics for their help with correspondence, copying, and the like and for their patience. Special thanks for all of this are due Elnora Ayers, Ann Gottorff and Shirley Zavaglia.

The staff of Academic Press was very helpful, patient, and supportive during the final stages of preparation of the manuscript. A very special "thank you" is due to Publisher Klaus Peters, Assistant Editor Susan Gay, Managing Editor Natasha Sabath and Production Editor Elizabeth Tustian.

Special thanks are also due to Ingram Olkin and Gerald J. Lieberman, the editors of the <u>Statistical</u> <u>Modeling</u> <u>and</u> <u>Decision</u> <u>Science</u> series. Their suggestions really helped make this a better book.

Marvin H.J.Gruber
Rochester, New York

PART I

INTRODUCTION AND MATHEMATICAL PRELIMINARIES

Chapter I

Introduction

1.0 Motivation for Writing This Book

During the past 15 to 20 years many different kinds of estimators have been proposed as alternatives to least squares estimators for the estimation of the parameters of a linear regression model. Ridge regression type estimators have proven to be very useful for multi-collinear data. As a result, since 1970, many papers and books have been written about ridge and related estimators.

There are a number of different mathematical formulations of ridge type estimators that are treated in the literature. Some of the formulations use Bayesian methods. Others employ methods within the context of the frequentist point of view. These include descriptions of the estimators as:

1. a special case of the Bayes estimator (BE);
2. a special case of the mixed estimator;
3. special cases of minimax estimators;
4. the solution to the problem of finding the point on an ellipsoid that is closest to a point in the parameter space.

These different approaches yield estimators with similar mathematical forms. Furthermore, for certain kinds of prior information and design matrices the estimators are the same. Throughout the literature, in the author's experience, most articles or books describe the estimators from only one of the above-mentioned points of view. For this reason the author felt that a comparative study of the similarities and differences amongst ridge type estimators obtained by the different methods should be undertaken.

1.1 Purpose of This Book

The objectives of this book are:

1. to give different mathematical formulations of ridge type estimators from both the frequentist and Bayesian points of view;
2. to explore the relationship between different kinds of prior information, the estimators obtained and their efficiencies;
3. to see how the methodology in points 1 and 2 can be used in a few special situations. (Of particular interest will be the Kalman filter and experimental design models.)

The rest of this chapter will contain:

1. a brief historical survey of work done by this and other authors (This survey is by no means exhaustive.);
2. some of the basic ideas about ridge and least square estimators;
3. an outline of the structure of the rest of the book.

1.2 Least Square Estimators and the Need for Alternatives

The method of least squares is often used by statisticians to estimate the parameters of a linear regression model. It consists of minimizing the sum of the squares of the differences between the predicted and the observed values of the dependent variables. As will be seen, when there appears to be a linear dependence between the independent variables, the precision of the estimates may be very poor. Ridge type estimators tend to be more precise for this situation. The derivation of the least square estimator and the derivation of the well-known Hoerl and Kennard ridge regression estimators will now be given.

Let

$$Y = X\beta + \varepsilon \qquad (1.2.1)$$

be a linear model. The matrix X is a known $n \times m$ matrix of rank $s \leq m$, β is an m dimensional vector and Y and ε are n dimensional random variables. Then ε has mean 0 and dispersion $\sigma^2 I$.

The least square (LS) estimator is derived by minimizing the sum of the squares of the error terms. Thus, minimize

$$\begin{aligned} F(\beta) &= (Y - X\beta)'(Y - X\beta) \\ &= Y'Y + \beta'X'X\beta - 2\beta'X'Y \end{aligned} \qquad (1.2.2)$$

by differentiation. Then

$$\frac{\partial F(\beta)}{\partial \beta} = X'X\beta - X'Y. \qquad (1.2.3)$$

Set (1.2.3) equal to zero. Thus, the normal equation

$$X'X\beta = X'Y \qquad (1.2.4)$$

is obtained. The solutions to the normal equation (1.2.4) are the least square estimators.

Sometimes in practice there may be a linear dependence between the m independent variables represented by the X matrix. Two situations may occur:

1. The case of exact multicollinearity: The rank of X is strictly less than s.
2. The case of near multicollinearity: The $X'X$ matrix has at least one very small eigenvalue. Then $\det X'X$ is very close to zero.

When X is of full rank the solution to (1.2.4) is

$$b = (X'X)^{-1}X'Y. \qquad (1.2.5)$$

When X is of less than full rank the solutions to the normal equations can still be found by finding a matrix G where

$$b = GX'Y. \qquad (1.2.6)$$

The matrix G is not unique and is called a generalized inverse. Generalized inverses and their properties will be considered in Chapter II. Some of the properties of least square estimators for the less than full rank model will be given in Chapter III.

The dispersion of b is

$$D(b) = \sigma^2 (X'X)^{-1}. \tag{1.2.7}$$

The total variance is

$$\text{Tr } D(b) = \sigma^2 \text{Tr}(X'X)^{-1} = \sigma^2 \sum_{i=1}^{s} \frac{1}{\lambda_i}, \tag{1.2.8}$$

where the λ_i are the non-zero eigenvalues of $X'X$. From the form of (1.2.8) notice that the total variance would be severely inflated if one or more of the λ_i was very small.

Hoerl and Kennard (1970) suggested using an estimator of the form

$$\hat{\beta} = (X'X + kI)^{-1}X'Y, \tag{1.2.9}$$

with k a positive constant. Then the total variance of the ridge estimators of β is

$$\text{TV}(\hat{\beta}) = \sigma^2 \sum_{i=1}^{s} \frac{1}{\lambda_i + k}. \tag{1.2.10}$$

Clearly, (1.2.10) is less than (1.2.8).

The estimator (1.2.9) is called a ridge regression estimator. It is obtained as the solution to the problem of minimizing the distance $B'B$ of a vector of parameters B from the origin subject to the side condition

$$(B - b)'X'X(B - b) = \Phi_0. \tag{1.2.11}$$

This is equivalent to finding the point on the ellipsoid centered at the LS estimator b that is closest to the origin. This ellipsoid may be thought of as a ridge, hence the name ridge regression.

The problem is solved by the method of Lagrange multipliers. Differentiate

$$L = B'B + \frac{1}{k}[(B - b)'X'X(B - b) - \Phi_0] \tag{1.2.12}$$

to obtain

$$B + \frac{1}{k}X'X(B - b) = 0. \qquad (1.2.13)$$

Solve (1.2.12) for B to obtain (1.2.9).

Example 1.2.1. How Ridge Estimators Are More Precise.
Examples of multicollinear data may be found frequently in economics. The data below is taken from the Economic Report of the President(1988). It represents the relationship between the dependent variable,Y (personal consumption expenditures) in billions of dollars, and three other independent variables, X_1, X_2, and X_3. The variable X_1 represents the Gross National Product, X_2 represents Personal Income (in billions of dollars), and X_3 represents the total number of employed people in the civilian labor force (in thousands).

Table 1.2.1   Economic Data

OBS	YEAR	Y	X_1	X_2	X_3
1	1965	440.7	705.1	552.0	71088
2	1966	477.3	772.0	600.8	72895
3	1967	503.6	816.4	644.5	74372
4	1968	552.5	892.7	707.2	75920
5	1969	579.9	963.9	772.9	77902
6	1970	640.0	1015.5	831.8	78678
7	1971	691.6	1102.7	894.0	79367
8	1972	757.6	1212.8	981.6	82153
9	1973	837.2	1359.3	1101.7	85064
10	1974	916.5	1472.8	1210.1	86794
11	1975	1012.8	1598.4	1313.4	85846
12	1976	1129.3	1782.8	1451.4	88752
13	1977	1257.2	1990.5	1607.5	92017
14	1978	1403.5	2249.7	1812.4	96048
15	1979	1566.8	2508.2	2034.0	98824
16	1980	1732.6	2732.0	2258.5	99303
17	1981	1915.1	3052.6	2520.9	100397
18	1982	2050.7	3166.0	2670.8	99526
19	1983	2234.5	3405.7	2838.6	100834
20	1984	2430.5	3772.2	3108.7	105005

Then, the matrix

$$X'X = \begin{bmatrix} 20 & 36571.3 & 29912.8 & 1750785 \\ 36571.3 & 84902181 & 69837276 & 3395418170 \\ 29912.8 & 69837276 & 57459487 & 2781058625 \\ 1750785 & 3395418170 & 2781058625 & 155483089935 \end{bmatrix}$$

has eigenvalues

$$\lambda_1 = 1.5561 \times 10^{11},$$
$$\lambda_2 = 1.8451 \times 10^7,$$
$$\lambda_3 = 3989.7,$$
$$\lambda_4 = .025724.$$

Notice that λ_4 is much smaller than the other three eigenvalues. Observe that

$$\sigma^2 \ \mathrm{Tr}(X'X)^{-1} = 38.874\sigma^2.$$

Most of the contribution comes from the fourth eigenvalue. When $k = 1$ the total variance of the ridge estimator is

$$\mathrm{TV}(\hat{\beta}) = \sigma^2(.975).$$

The total variance is thus reduced by a factor of almost 40, a substantial reduction.

Example 1.2.2. Least Square and Ridge Regression Estimators for Economic Data. Using the data in Example 1.2.1 the solution to the normal equation

$$X'X\hat{\beta} = X'Y$$

is

$$\hat{\beta} = (X'X)^{-1}X'Y,$$

where

$(X'X)^{-1}$

$$= \begin{bmatrix} 38.874 & 0.0279 & -0.0258 & -0.0006 \\ 0.0279 & 0.0001 & -0.0001 & -4.957 \times 10^{-7} \\ -0.0258 & -0.0001 & 0.0002 & 4.763 \times 10^{-7} \\ -0.0006 & -4.957 \times 10^{-7} & 4.763 \times 10^{-7} & 8.893 \times 10^{-9} \end{bmatrix}$$

and

$$(X'Y) = \begin{bmatrix} 23129.9 \\ 54034043 \\ 44456444 \\ 2150387618 \end{bmatrix}.$$

Thus,

$$\begin{bmatrix} \hat{\beta}_0 \\ \hat{\beta}_1 \\ \hat{\beta}_2 \\ \hat{\beta}_3 \end{bmatrix} = \begin{bmatrix} 265.341 \\ 0.38491 \\ 0.3708 \\ -0.0042 \end{bmatrix}.$$

Thus, the fitted equation is

$$Y = 265.341 + .3849X_1 + .3708X_2 - .0042X_3.$$

When $k = 1$ the ridge regression estimator is

$$
\begin{bmatrix} \hat{\beta}_0 \\ \hat{\beta}_1 \\ \hat{\beta}_2 \\ \hat{\beta}_3 \end{bmatrix} = \begin{bmatrix} 5.17911 \\ 0.19938 \\ 0.54313 \\ -0.0031 \end{bmatrix}.
$$

Exercise 1.2.1. For a linear model $Y = X\beta + \varepsilon$ find the least square estimators of β when

A. $Y = \begin{bmatrix} 1.2 \\ 1.5 \\ 1.4 \\ 1.7 \\ 1.9 \end{bmatrix}$, $X = \begin{bmatrix} 1 & 3.4 & 2.1 \\ 1 & 2.8 & 2.3 \\ 1 & 1.9 & 1.1 \\ 1 & 6.2 & 1.4 \\ 1 & 4.1 & 1.6 \end{bmatrix}$;

B. $Y = \begin{bmatrix} 4.2 \\ 4.5 \\ 3.7 \\ 4.8 \\ 2.6 \end{bmatrix}$, $X = \begin{bmatrix} 1 & 1.1 & 2.0 \\ 1 & 2.1 & 4.3 \\ 1 & 3.4 & 6.7 \\ 1 & 2.2 & 4.5 \\ 1 & 1.4 & 2.7 \end{bmatrix}$.

Exercise 1.2.2. For each model in Exercise 1.2.1 above find the ridge regression estimator when $k = .1, .5, 1, 2, 10$.

It will be shown in Chapter V that the estimator (1.2.9) is biased while the least square estimator is unbiased. Hoerl and Kennard

(1970) showed that there exists a k such that the total Mean Square Error (MSE)

$$M = E(\hat{\beta} - \beta)'(\hat{\beta} - \beta) \qquad (1.2.14)$$

of a ridge estimator is smaller than that of a LS estimator. It has already been pointed out that the ridge estimator has a smaller total variance than the LS estimator.

The proper choice of k to obtain a small MSE is a question that has received much attention in the literature. Some analytical work and numerous simulation studies have been done. Two of the better known simulation studies are Hoerl, Kennard and Baldwin (1975) and Dempster et al.(1977).

It will be shown how the ridge regression estimator and other estimators with similar mathematical forms are special cases of Bayes, minimax, generalized ridge and mixed estimators in the chapters that follow.

1.3 Historical Survey

The Bayes, mixed, ridge and minimax estimators were discovered by different authors quite independently. A careful study of the literature shows that these estimators have many similarities. Moreover these estimators frequently have smaller mean square error and are more precise as estimators of the parameters in a linear regression model than the least square estimator. This was found to be very helpful in the analysis of multicollinear data.

The following is a brief review of some of the important literature on the estimators to be considered in this book.

Stein (1956) showed that the maximum likelihood estimator for the mean vector in a multivariate normal distribution with at least 3 variates is inadmissible. That means that for certain values in the parameter space other estimators existed that had a smaller risk or mean square error. An explicit estimator with that property is given by James and Stein (1961). It appears that the work of James and Stein was a starting point for the study of alternatives to least square estimators.

The mixed estimator was first described by Theil and Goldberger

(1961). The mixed estimator is a LS estimator for a linear model augmented by taking additional observations.

A general discussion of the multicollinearity problem is given by Silvey (1969) using eigenvalues of the design matrix and the singular value decomposition.

The work that is probably the most frequently cited in the literature of ridge regression is Hoerl and Kennard's paper (1970a). The ridge regression estimator is obtained as the estimator whose distance to an ellipsoid centered at the LS estimator from the origin of a parameter space is a minimum. Their paper shows the MSE optimality over least squares, the connection with Stein estimators and the Bayesian formulation. The methods are applied to several data sets in Hoerl and Kennard (1970b).

Marquardt (1970) describes some properties of the LS estimator for the non-full rank model that is derived by employing the Moore-Penrose generalized inverse. He notes that the ridge regression estimator of Hoerl and Kennard may also be viewed as a LS estimator where additional "fictitious observations" are taken.

Banerjee and Karr (1971) also characterize the ridge regression estimator as an LS estimator with respect to an augmented linear model. For this characterization the estimator is shown to be unbiased.

Duncan and Horn (1972) showed how the Gauss-Markov theorem could be extended to include Bayes and Ridge type estimators. More work on this topic was done by Harville (1976). A critique of their methods and those of other authors was given by Pfeffermann (1984).

Mayer and Willke (1973) obtains a contraction estimator by replacing the distance function in Hoerl and Kennard (1970) by one weighted by the design matrix. They propose use of the contraction estimator as an alternative to the LS estimator.

Goldstein and Smith (1974) define a class of shrinkage estimators. They reparametize the linear model so that the design matrix is diagonal. For this reparametized model they show that for certain values of the ridge parameters the individual MSE of components of a shrinkage estimator is less than that of components of an LS estimator. They obtain the optimum ridge parameters for the generalized ridge regression estimator. They also have a Bayesian formulation of ridge estimators and show how it approximates the LS estimator obtained using the Moore-Penrose inverse.

Lowerre (1974) gives conditions for the MSE of biased ridge type estimators to be smaller than that of the LS estimator. He considers both the full and the non-full rank case.

Many of the comparisons of ridge regression estimators were accomplished by means of simulation studies. One such study was that of Hoerl, Kennard and Baldwin (1975). This dealt primarily with the ridge estimator. Another such study was that of Dempster, Schatzoff and Wermuth (1977). It made comparisons of various ridge, James-Stein and other estimators.

C.R. Rao (1973, 1975) derives the linear Bayes estimator. For the full rank case he obtains alternative forms. The ridge and contraction estimators turn out to be special cases of the Bayes estimator (BE). C.R. Rao (1975) considers the problem of simultaneous estimation of the parameters for linear models employing empirical Bayes estimators.

In C.R. Rao (1976) the concept of an admissible linear estimator is taken up. It is shown that the admissible linear estimators are either Bayes linear estimators or limits of Bayes linear estimators. A brief presentation of the ridge and minimax estimators as a special case of the Bayes linear estimators is given there. Zontek (1987) extends Rao's (1976) result.

Later on in the text (Chapter IV) it will be explained how ridge type estimators may be represented as LS estimators multiplied by an appropriate matrix often referred to in the literature as a shrinkage factor. The shrinkage is either toward zero or a nonzero mean vector based on prior information. Swindel (1976) considers properties of ridge estimators for shrinkage toward a mean vector other than zero. This work also gives some necessary conditions for the mean square error of parametric functions for ridge estimators to be less than that of LS estimators.

Two brief notes by Farebrother (1976, 1978) give necessary and sufficient conditions for ridge or general shrinkage type estimators to have a smaller MSE than LS estimators for the parametric functions.

Brown (1978) studies ridge estimation for the non-full rank model. A characterization of the LS estimator as a limit of ridge regression estimators is given.

A review article by Vinod (1978) includes a discussion of ridge and

James-Stein estimators from a Bayesian point of view. This article has an extensive bibliography.

Bayes and empirical Bayes estimators are formulated in Gruber (1979) for estimable functions of the parameters of non-full rank models. Ridge and contraction estimators are shown by him to be special cases of Bayes estimators. The mean square errors of Bayes estimators for estimable functions are compared with those of the least square estimators. Some general results for comparison of the prediction error of two Bayes estimators are obtained.

D. Trenkler and G. Trenkler (1980) present a general criterion of comparison of biased estimators that contains the results of Farebrother (1976) as a special case.

Wax and Haitovsky (1980) generalized the results and methods of Hoerl and Kennard to a class of ridge type estimators. They also explained how a generalized ridge regression estimator is a BE or mixed estimator.

The comparison of the MSE of minimax estimators and mixed estimators for a quadratic loss function is undertaken by Teräsvirta (1981). Teräsvirta obtains necessary and sufficient conditions for minimax estimators to have a smaller MSE than the mixed estimators. He also compares the performance of the mixed and minimax estimators with that of the LS estimator.

Wang (1982) presents biased estimators for the non-full rank linear model and explains how ridge regression estimators are special cases of BE.

Results similar to those of Wang (1982) are obtained independently by Gruber and P.S.R.S. Rao (1982). In addition they obtain alternative forms of the BE for the non-full rank model and give criteria on the prior distribution for one BE to have a smaller average MSE than another. This will be taken up in Chapters III, IV, and VI.

Peele and Ryan (1982) observe that minimax estimators have a smaller MSE than LS estimators. They also explain how ridge regression estimators are special cases of the minimax estimator.

Price (1982) observes that comparisons are possible for the MSE of parametric functions when ridge and LS estimators are being compared but that it is not always possible to compare two ridge type estimators using this criterion.

Toutenburg's book (1982) considers both minimax and mixed estimation under incorrect prior information. The relationship between ridge, minimax, and mixed estimators is explained. This work is from a purely non-Bayesian point of view. Our work will compare the results of both Bayesian and non-Bayesian methods. A lengthy bibliography appears at the end of Toutenburg's book.

The results of Farebrother (1976, 1978) are generalized in Farebrother (1984) to compare the MSE of a generalized ridge estimator with that of a restricted LS estimator.

The Theil-Goldberger mixed estimator is shown to be a special type of Bayes estimator by Saxena (1984). A somewhat different derivation of the same result (to be described in Chapter IV) was obtained by Gruber (1985). The problem of incorrect prior information for Bayes estimators and mixed estimators for non-full rank linear models to be taken up in Chapter VIII is also described there.

Swamy et al. (1985) observe that the mixed estimator and the Bayes estimator are not equivalent when the prior dispersion is of non-full rank. They also present a technique for the measurement of multicollinearity.

The problem of robustness of prior information is taken up by Berger and Berliner (1986). In this context they discuss ε contaminated prior distributions.

Teräsvirta (1986) gives necessary and sufficient conditions for the comparison of two heterogeneous linear estimators for both parametric functions (matrix loss) and for predictive loss functions. His results could be applied to the comparison of Bayes, minimax or mixed estimators. However, the geometric form of the region where one estimator is "better" than another can be quite complicated. It would not be an ellipsoid as is the case when ridge and LS estimators are compared.

In Toutenburg (1986) weighted mixed regression is discussed. Here the auxiliary information is weighted by an appropriate scalar weight between zero and one. Comparisons of the MSE with that of the LS estimator are given. The methods of mixed estimation are applied to the missing value problem.

Bunke and Bunke (1986) contains a discussion of ridge, minimax and James-Stein type estimators. This book is an English version of Humak (1977). The mathematical level at which the various topics are presented is quite sophisticated. At the end of Chapters 1–4 and 7 of Bunke and Bunke the reader will find good bibliographic references.

Another good source for references is Berger (1985). This book has a good treatment of Bayes, empirical Bayes estimators and minimax estimators. The subject of Bayesian robustness is touched upon. (This topic will be taken up for linear models in Chapter VIII of this book.)

Pliskin (1987) considers a modified ridge regression estimator proposed by Swindel (1976) based on a prior mean. He obtains a necessary and sufficient condition for Swindel's estimator to have a smaller risk than the usual ridge estimator when both estimators are computed using the same value of k. This problem is further studied by Trenkler (1988) for the case of different ridge parameters.

Lauterbach and Stahlecker (1987, 1988) consider an ordinary linear regression model whose parameters are constrained to a given ellipsoid. They obtain a sequence of approximate minimax estimators with an exact minimax estimator as a cluster point. This minimax estimator is compared to the the LS estimator via a simulation study. The situation considered in their papers is somewhat more general than that considered in this book.

Stahlecker and Trenkler (1988) develop minimax estimators for a linear regression model where the parameters are simultaneously constrained to an ellipsoid and by exact linear restrictions.

The comparison of the MSE between restricted least squares, mixed and weighted least squares is taken up in Toutenburg (1988). In this paper nested restrictions are also considered.

Schipp et al. (1988) consider a situation where the parameter vectors are constrained by linear equations and lie in a given ellipsoid. They show that when the weight matrix A of the risk function has rank one the restricted minimax estimator combining both types of prior information exists. For general NND matrices A other estimators are proposed and compared to the LS estimator via a simulation study.

The Jackknife estimator was introduced by Quenouille (1956) as a technique for reduction of bias. The ridge regression type estimators are biased. Nyquist (1988) shows how to reduce the bias of ridge regression estimators using a Jackknife estimator. The properties of the Jackknife ridge estimator are also studied by Singh, Chaubey and Dwivedi (1986) and Nomura (1988). An example of this procedure is given in Exercise 7.1.4.

Chawla (1988) obtains necessary and sufficient conditions for the matrix mean square error of a generalized ridge estimator to be smaller than that of an LS estimator. Results of this type obtained earlier by the author are taken up in Section 7.3.

The results for comparison of two heterogeneous linear estimators of Teräsvirta (1986) are applied to the comparison of two mixed estimators in Teräsvirta (1988).

Liski (1988) develops criteria for comparison of linear admissible estimators, including ridge type estimators, with LS estimators similar to those of aforementioned authors. He develops a test procedure for chosing a linear admissible estimator in preference to a LS estimator.

The list of references whose contents are described above is far from exhaustive. However, the interested reader can use them together with their bibliographies to find many entry points to the literature.

1.4 The Structure of the Book

In order that this book be reasonably self-contained, some mathematical and statistical preliminaries are given in Chapter II. In 2.1 the matrix theory used in the book that is not usually covered in a first course is summarized. This includes positive definiteness of a matrix, the singular value decomposition and the important properties of generalized inverses.

Section 2.2 explains what the Bayes estimator is and describes some of its properties. Section 2.3 describes the minimax estimator.

Section 2.4 gives a result in the literature due to Theobald, which is important to the latter part of the book.

Chapter III explains how the least square estimator, the generalized ridge estimator, the BE and the mixed and minimax estimators are solutions to different optimization problems with certain common elements.

Chapter IV considers the relationships between the different estimators, e.g., when the different estimators are either equivalent or expressible in terms of one another.

Chapter V defines the measures of efficiency used to compare the estimators.

The average MSE of the BE, the MSE of the mixed estimator and the optimal values of the optimization criteria are considered in Chapter VI. Comparisons of the average MSE of different estimators are made.

The conditional MSE of the BE and the corresponding MSE neglecting prior assumptions of the mixed estimator is the subject of Chapter VII.

Chapter VIII considers the problem of robustness of prior information. The properties of the MSE are considered there when the prior averaged over is not the prior distribution that was used to derive the estimators.

Chapter IX applies the methods of the previous chapters to the discrete Kalman filter. Chapter X does the same for simple experimental design models.

The book may be thought of as having four parts:

I. Introduction and Mathematical Preliminaries
   (Chapters I and II).
II. The Estimators, Their Derivations and the Relationship between Them
   (Chapters III and IV).
III. The Comparison of Efficiences of Estimators
   (Chapters V−VIII).
IV. Applications
   (Chapters IX and X).

Chapter II

Mathematical and Statistical Preliminaries

2.0 Introduction

The purpose of this chapter is to summarize some basic ideas from matrix theory and statistics that are used throughout this monograph. In the author's experience the topics to be discussed here are sometimes omitted from undergraduate courses. Readers with a good background in these areas may proceed to Chapter III and refer back to this chapter as nececessary.

In comparing the efficiencies of estimators the properties of positive semidefinite matrices prove to be quite useful. The singular value decomposition is useful to the study of generalized inverses and in establishing conditions for equivalences of the estimators to be studied. Generalized inverses help in finding solutions to normal equations for least square estimators when the design matrix is of non-full rank. These topics are taken up in Section 2.1.

Section 2.2 presents the Bayes estimator as the expected value of a posterior distribution that is obtained by taking prior information about a parameter together with a likelihood function obtained from sample observations and applying Bayes theorem. The resulting posterior distribution of the parameter, obtained using both sample and prior information, is an "after the fact" distribution. The Bayes estimator is optimum because it is the estimator with the smallest

19

average mean square error. It is also admissible; that means that no other estimator has a smaller MSE for any parameter values. In other words, no other estimator is uniformly better i.e. better for all parameter values (An estimator is considered to be "better" if it has smaller MSE.).

The minimax estimator is described in Section 2.3. The minimax estimator is the one whose maximum MSE on a certain ellipsoid is minimized.

A criterion for comparison of estimators under different loss functions (Theobald's result) is contained in Section 2.4.

Section 2.5 contains some useful inequalities that will later enable comparison of different BE and least square estimators.

Section 2.6 contains some results on the Kronecker product and patterned matrices. The Kronecker product gives useful representation to sums of squares and matrices used in ANOVA. The results on patterned matrices will be used in Chapter X.

2.1 Matrix Theory Results

2.1.1 Positive Semidefinite Matrices

The results, definitions and concepts from matrix theory stated below will be used extensively in this monograph.

An $m \times m$ matrix A is positive semidefinite or non-negative definite NND if and only if (iff) for every m dimensional column vector p the quadratic form

$$p'Ap \geq 0 \qquad\qquad (2.1.1)$$

and A is symmetric $(A = A')$. (A' means the transpose of A.) If in addition to (2.1.1) $p'Ap = 0$ iff $p = 0$ or equivalently

$$p'Ap > 0 \qquad\qquad (2.1.2)$$

for every non-zero m dimensional vector, the matrix A will be said to be positive definite (PD).

Some important frequently used properties of PD matrices are:

1. All of their eigenvalues are positive. (In case of positive semidefinite matrices the eigenvalues are non-negative.)

2. The determinants of the principal minors are positive.

For an $n \times n$ symmetric matrix the principal minors are the $k \times k$ submatrices $1 \leq k \leq n$ containing the diagonal elements a_{ii} $1 \leq i \leq k$. Thus, if

$$A = \begin{bmatrix} a_{11} & a_{12} & \cdots & a_{1n} \\ a_{12} & a_{22} & \cdots & a_{2n} \\ \vdots & \vdots & \ddots & \vdots \\ a_{1n} & a_{2n} & \cdots & a_{nn} \end{bmatrix},$$

then

$$a_{11} > 0, \quad \det \begin{bmatrix} a_{11} & a_{12} \\ a_{12} & a_{22} \end{bmatrix} > 0,$$

$$\det \begin{bmatrix} a_{11} & a_{12} & a_{13} \\ a_{12} & a_{22} & a_{23} \\ a_{13} & a_{23} & a_{33} \end{bmatrix} > 0, \ldots, \det A > 0. \tag{2.1.3}$$

Example 2.1.1. A Positive Definite Matrix. Let

$$A = \begin{bmatrix} 1 & 1 & 0 \\ 1 & 4 & 0 \\ 0 & 0 & 5 \end{bmatrix}.$$

Notice that the principal minors are

$$A_{11} = \begin{bmatrix} 1 & 1 \\ 1 & 4 \end{bmatrix} = 3 > 0 \quad \text{and} \quad \det(A) = 15.$$

Thus, the principal minors are PD. Thus, A is PD.

The eigenvalues of A are given by

$$\lambda_1 = 5.000, \quad \lambda_2 = 4.303, \quad \lambda_3 = 0.697.$$

They are all positive numbers.

Exercise 2.1.1. Which of the following matrices are (1) positive definite? (2) positive semidefinite?

A. $\begin{bmatrix} 1 & 2 \\ 2 & 4 \end{bmatrix}$ B. $\begin{bmatrix} 5 & 12 \\ 12 & 100 \end{bmatrix}$ C. $\begin{bmatrix} \frac{5}{2} & -\frac{1}{2} \\ -\frac{1}{2} & \frac{5}{2} \end{bmatrix}$

D. $\begin{bmatrix} \frac{3}{2} & -\frac{3}{2} \\ -\frac{3}{2} & \frac{3}{2} \end{bmatrix}$ E. $\begin{bmatrix} 1 & 1 & 2 \\ 1 & 2 & 1 \\ 1 & 1 & 3 \end{bmatrix}$ F. $\begin{bmatrix} 11 & -7 & -4 \\ -7 & 11 & -4 \\ -4 & -4 & 8 \end{bmatrix}$

G. $\begin{bmatrix} 1 & -3 \\ -3 & 1 \end{bmatrix}$.

In the sequel non-negative definite matrices will frequently represent dispersions or mean square errors of estimators. The quadratic forms will represent the variance or in some cases the mean square error of parametric functions of estimators $p'\beta$. A means of comparing the "size" of these quantities is needed. A non-negative definite NND matrix A will be said to be greater than or equal to B if $A - B$ is non-negative definite. A matrix A will be considered greater than B if $A - B$ is positive definite. If should be observed that two NND matrices need not be greater than or less than one another.

In set theory a partially ordered set is one where for elements a, b, c
(1) $a \le a$
(2) $a \le b, b \le a$ implies $a = b$
(3) $a \le b, b \le c$ implies $a \le c$
A totally ordered set satisfies (1)–(3) above and for any two elements $a, b, a \le b$ or $b \le a$. Thus, the NND matrices are partially but not totally ordered.

Example 2.1.2. Order Relationships between Matrices. Let

$$A = \begin{bmatrix} 9 & 2 \\ 2 & 6 \end{bmatrix}, \ B = \begin{bmatrix} 5 & 2 \\ 2 & 3 \end{bmatrix} \text{ and } C = \begin{bmatrix} 9 & 4 \\ 2 & 3 \end{bmatrix}.$$

Then

$$A - B = \begin{bmatrix} 4 & 0 \\ 0 & 3 \end{bmatrix}$$

is positive definite and

$$A - C = \begin{bmatrix} 0 & 0 \\ 0 & 3 \end{bmatrix}$$

is positive semidefinite. Thus, $A > B$ and $A \geq C$. However, Example 2.1.3 below presents two matrices whose difference is not positive semidefinite.

Example 2.1.3. Two Matrices Whose Difference Is Not Positive Semidefinite.

$$b = \begin{bmatrix} 4 \\ 3 \end{bmatrix}, \quad c = \begin{bmatrix} 1 \\ 2 \end{bmatrix}$$

Now

$$bb' = \begin{bmatrix} 4 \\ 3 \end{bmatrix} [4\ 3] = \begin{bmatrix} 16 & 12 \\ 12 & 9 \end{bmatrix},$$

$$cc' = \begin{bmatrix} 1 \\ 2 \end{bmatrix} [1\ 2] = \begin{bmatrix} 1 & 2 \\ 2 & 4 \end{bmatrix}.$$

Both bb' and cc' are positive semidefinite. However for

$$D = bb' - cc' = \begin{bmatrix} 15 & 10 \\ 10 & 5 \end{bmatrix},$$

$$\det(D) = -25.$$

Thus, D is not positive semidefinite and bb' is not greater than or equal to cc'.

Exercise 2.1.2. For each of the following pairs of matrices determine whether
(1) $A > B$ (2) $A \geq B$ (3) $A < B$ (4) $A \leq B$
or (5) none of the relationships in (1)−(4) hold. Also find the eigenvalues of $A - B$.

A. $A = \begin{bmatrix} 2 & 1 \\ 1 & 2 \end{bmatrix}$, $B = \begin{bmatrix} 3 & 1 \\ 1 & 2 \end{bmatrix}$.

B. $A = aa'$, $B = bb'$ where

(1) $a = \begin{bmatrix} 1 \\ 2 \\ 3 \end{bmatrix}$, $b = \begin{bmatrix} 2 \\ 5 \\ 10 \end{bmatrix}$;

(2) $a = \begin{bmatrix} 1 \\ 2 \\ 3 \end{bmatrix}$, $b = \begin{bmatrix} 2 \\ 4 \\ 6 \end{bmatrix}$;

C. $A = 18I$, $B = \begin{bmatrix} 11 & -7 & -4 \\ -7 & 11 & -4 \\ -4 & -4 & 8 \end{bmatrix}$.

D. $A = 10I$, B as in part C.

E. $A = 20I$, B as in part C.

2.1.2 The Singular Value Decomposition

A simple and useful representation of a matrix in terms of its eigenvalues and orthogonal eigenvectors is the singular value decomposition (SVD). The SVD is quite useful in the study of generalized inverses and in establishing algebraic equivalence of different estimators as will be seen in Chapter IV.

An $m \times m$ matrix C is orthogonal if

$$C'C = I \text{ and } CC' = I. \qquad (2.1.4)$$

Let X be an $n \times m$ matrix of rank s. Then the singular value decomposition of X is

$$X = [S' \ T'] \begin{bmatrix} \Lambda^{1/2} & 0 \\ 0 & 0 \end{bmatrix} \begin{bmatrix} U' \\ V' \end{bmatrix} = S'\Lambda^{1/2}U', \qquad (2.1.5)$$

where $[S' \ T']$ and $[U \ V]$ are orthogonal matrices, U is $m \times s$, V is $m \times (m - s)$, S' is $n \times s$, T' is $n \times (n - s)$ and Λ is the $s \times s$ diagonal matrix of singular values (eigenvalues) λ_i, $1 \leq i \leq s$ where $\lambda_1 \geq \lambda_2 \geq \cdots \geq \lambda_s$.

Notice that, from (2.1.5)

$$X'X = U\Lambda^{1/2}SS'\Lambda^{1/2}U' = U\Lambda U'. \qquad (2.1.6)$$

Furthermore

$$XUU' = S'\Lambda^{1/2}U'UU' = S'\Lambda^{1/2}U' = X \qquad (2.1.7)$$

and

$$XV = S'\Lambda^{1/2}U'V = 0. \qquad (2.1.8)$$

Example 2.1.4. The SVD of a Matrix. Let

$$X = \begin{bmatrix} 1 & 1 & 0 \\ \frac{1}{\sqrt{2}} & -\frac{1}{\sqrt{2}} & 0 \\ 0 & 0 & 0 \end{bmatrix}.$$

Then

$$X'X = \begin{bmatrix} \frac{3}{2} & \frac{1}{2} & 0 \\ \frac{1}{2} & \frac{3}{2} & 0 \\ 0 & 0 & 0 \end{bmatrix}$$

and

$$XX' = \begin{bmatrix} 2 & 0 & 0 \\ 0 & 1 & 0 \\ 0 & 0 & 0 \end{bmatrix}.$$

The reader can easily verify that the eigenvalues of $X'X$ and XX' are 2, 1 and 0. Then the SVD of

$$X = \begin{bmatrix} 1 & 0 & 0 \\ 0 & 1 & 0 \\ 0 & 0 & 1 \end{bmatrix} \begin{bmatrix} \sqrt{2} & 0 & 0 \\ 0 & 1 & 0 \\ 0 & 0 & 0 \end{bmatrix} \begin{bmatrix} \frac{1}{\sqrt{2}} & \frac{1}{\sqrt{2}} & 0 \\ \frac{1}{\sqrt{2}} & -\frac{1}{\sqrt{2}} & 0 \\ 0 & 0 & 1 \end{bmatrix}.$$

Example 2.1.5. Another SVD. Let $X'X$ be as in Example 1.2.1. From (2.1.6),

$$X'X = U\Lambda U',$$

where (rounded to five decimal places)

$$U = \begin{bmatrix} 0.00001 & 0.00012 & 0.00097 & 1.00000 \\ 0.02183 & -0.76275 & -0.64632 & 0.00072 \\ 0.01788 & -0.64607 & 0.76307 & -0.00066 \\ 0.99960 & 0.02822 & 0.00047 & -0.00002 \end{bmatrix}$$

and

$$\Lambda = \begin{bmatrix} 1.5561 \times 10^{11} & 0 & 0 & 0 \\ 0 & 1.8451 \times 10^7 & 0 & 0 \\ 0 & 0 & 3989.7 & 0 \\ 0 & 0 & 0 & .025724 \end{bmatrix}.$$

Example 2.1.6. Another SVD. Let

$$X = \begin{bmatrix} \frac{4}{\sqrt{2}} & \frac{2}{\sqrt{2}} \\ -\frac{4}{\sqrt{2}} & \frac{2}{\sqrt{2}} \end{bmatrix}.$$

Then

$$X'X = \begin{bmatrix} 16 & 0 \\ 0 & 4 \end{bmatrix} \text{ and } XX' = \begin{bmatrix} 10 & -6 \\ -6 & 10 \end{bmatrix}.$$

Both $X'X$ and XX' have eigenvalues 16 and 4. Observe that for $\lambda = 16$

$$\begin{bmatrix} -6 & -6 \\ -6 & -6 \end{bmatrix} \begin{bmatrix} u_1 \\ u_2 \end{bmatrix} = 0$$

has a solution $u_1 = \frac{1}{\sqrt{2}}$ and $u_2 = -\frac{1}{\sqrt{2}}$. For $\lambda = 4$

$$\begin{bmatrix} 6 & -6 \\ -6 & 6 \end{bmatrix} \begin{bmatrix} v_1 \\ v_2 \end{bmatrix} = \begin{bmatrix} 0 \\ 0 \end{bmatrix}$$

has a solution $v_1 = \frac{1}{\sqrt{2}}$ and $v_2 = \frac{1}{\sqrt{2}}$. Then the SVD of

$$X = \begin{bmatrix} \frac{1}{\sqrt{2}} & \frac{1}{\sqrt{2}} \\ -\frac{1}{\sqrt{2}} & \frac{1}{\sqrt{2}} \end{bmatrix} \begin{bmatrix} 4 & 0 \\ 0 & 2 \end{bmatrix} \begin{bmatrix} 1 & 0 \\ 0 & 1 \end{bmatrix}.$$

Exercise 2.1.3. Let

$$X = \begin{bmatrix} 1 & 1 & 0 \\ 1 & 1 & 0 \\ 1 & 0 & 1 \\ 1 & 0 & 1 \end{bmatrix}.$$

Find the singular value decomposition of
A. $X'X$;
B. XX';
C. X.

Exercise 2.1.4. Use the singular value decomposition to show that
A. If A is an $m \times m$ PD matrix then there is a nonsingular matrix M where $A = MM'$.
B. If A is an $m \times m$ positive semidefinite matrix of rank $s < m$ then there is an $m \times s$ matrix N such that $A = NN'$.
C. For parts A and B above find a matrix C where $A = C^2$ in terms of the SVD matrices.

2.1.3 Generalized Inverses

Frequently in the solution of regression problems it is necessary to solve a system of equations of the form

$$Ax = y \qquad (2.1.9a)$$

where x is an n dimensional vector and A is an $m \times n$ matrix of rank $s \leq m$, and y is an m dimensional vector.

Where $n = m$ and $s = m$ of course $x = A^{-1}y$. Assume that the system of equations in (2.1.9a) is consistent (has a solution). The form of the solution would be

$$x = Gy, \qquad\qquad (2.1.9b)$$

where G is an $m \times n$ matrix. The matrix G is called a generalized inverse. In general the solution (2.1.9b) need not be unique.

If A is a $n \times m$ matrix the totality of column vectors Ax for all m dimensional real vectors x can be shown to be a vector subspace of the vector space of m dimensional real column vectors. This vector subspace is denoted as $M(A)$ and is called the column space of A. A formal definition of generalized inverses is given in Definition 2.1.1 below.

Definition 2.1.1. Consider any $m \times n$ matrix A. Let x be an n dimensional vector. Let y be m dimensional vector. A generalized inverse of A is an $n \times m$ matrix G such $x = Gy$ is a solution to the equation $Ax = y$ for every $y \in M(A)$.

The following theorem leads to an alternative definition of generalized inverse.

Theorem 2.1.1. Let A be an $m \times n$ matrix. The $n \times m$ matrix G is a generalized inverse of A iff $AGA = A$.

Proof.

1) Let $y \in M(A)$. Then by Definition 2.1.1 $AGy = y$. Let $y = a_j \in M(A)$ where the a_j are columns of A. Then $AGa_i = a_i$ with $1 \leq i \leq n$. Thus $AGA = A$.

2) Suppose $AGA = A$. Let $y \in M(A)$. Then let $x = Gy$. It must be shown that $x = Gy$ is a solution to $Ax = y$. But $y \in M(A)$ so $AGy = AGAx = Ax = y$. Thus, for all $y \in M(A)$, Gy is a solution to $Ax = y$. ∎

Thus, a definition of a generalized inverse that is equivalent to Definition 2.1.1 above is:

Definition 2.1.2. G is a generalized inverse of A iff $AGA = A$.

Example 2.1.7. Generalized Inverses. Let

$$A = \begin{bmatrix} 1 & 1 & 1 \\ 1 & 2 & 3 \\ 2 & 4 & 6 \end{bmatrix}.$$

Two generalized inverses of A are

$$G_1 = \begin{bmatrix} 2 & -1 & 0 \\ -1 & 1 & 0 \\ 0 & 0 & 0 \end{bmatrix}$$

and

$$G_2 = \begin{bmatrix} 0 & 3 & -1 \\ 0 & -1 & 1 \\ 0 & 0 & 0 \end{bmatrix}.$$

Exercise 2.1.5. A. Show for matrices A, G_1 and G_2 in Example 2.1.7 that

$$AG_1 A = A \quad \text{and} \quad AG_2 A = A.$$

(In other words verify that G_1 and G_2 are both generalized inverses of A.)
B. Show that

$$G_1 AG_1 = G_1 \quad \text{and} \quad G_2 AG_2 = G_2.$$

Such generalized inverses are called reflexive generalized inverses.
C. Are $G_1 A$, AG_1, $G_2 A$ or AG_2 symmetric matrices?

The concept of Moore-Penrose inverse gives a mathematically convenient way to

1. represent generalized inverses (see Theorem 2.1.2 and corollaries);
2. obtain some of the algebraic results of the next chapter;
3. characterize the solution to least square equations;
4. view least square estimators as a limit of ridge estimators.

Definition 2.1.3. A Moore-Penrose inverse is a generalized inverse where

(1) $GAG = G$
(2) AG is a symmetric matrix
(3) GA is a symmetric matrix.

A generalized inverse of A is denoted by A^-. The Moore Penrose inverse of A is denoted by A^+. A matrix A can have many generalized inverses, but only one Moore-Penrose inverse. When A has full rank the unique ordinary inverse is the generalized inverse. Theorem 2.1.2 characterizes generalized inverses in terms of the Moore-Penrose inverse.

Theorem 2.1.2. Let X be an $n \times m$ matrix of rank s with SVD in (2.1.5). Then

1. $X^+ = U\Lambda^{-1/2}S$ (2.1.10a)
2. $U'X^-S' = \Lambda^{-1/2}$ for every generalized inverse of X.   (2.2.10b)
3. A representation of X^- is

$$X^- = [U\ V]\begin{bmatrix} \Lambda^{-1/2} & A_1 \\ A_2 & A_3 \end{bmatrix}\begin{bmatrix} S \\ T \end{bmatrix}$$

$$(2.1.11)$$

$$= X^+ + VA_2S + UA_1T + VA_3T,$$

with $A_1 = U'X^-T'$, $A_2 = V'X^-S'$ and $A_3 = V'X^-T'$.
4. If C_1, C_2 are arbitrary matrices and $C_3 = C_2 \wedge C_1$, then

$$M = [U\ V]\begin{bmatrix} \Lambda^{-1/2} & C_1 \\ C_2 & C_3 \end{bmatrix}\begin{bmatrix} S \\ T \end{bmatrix}$$

$$(2.1.12)$$

is a generalized inverse of X.

Some or all of conditions (2)–(4) in Definition 2.1.3 are satisfied by special forms of M in (2.1.12). If $C_1 = 0$ then condition (3) is satisfied. If $C_2 = 0$ then condition (4) is satisfied. Furthermore if $C_2\Lambda^{1/2}C_1 = C_3$ in (2.1.12) condition (2) holds. All four conditions are satisfied when $C_1 = C_2 = C_3 = 0$. Then M is a Moore-Penrose inverse of X.

The proof of Theorem 2.1.2 is below.

1. Do a straightforward check of the axioms in Definitions 2.1.2 and 2.1.3. To verify (1), (2) and (3) observe that
$$XX^+X = S'\Lambda^{1/2}U'U\Lambda^{-1/2}SS'\Lambda^{1/2}U' = S\Lambda^{1/2}U' = X, \quad X^+XX^+$$
$$= U\Lambda^{-1/2}SS'\Lambda^{1/2}U'U\Lambda^{-1/2}S = U\Lambda^{-1/2}S = X^+ \text{ and}$$
$$X^+X = U\Lambda^{-1/2}SS'\Lambda^{1/2}U' = UU'.$$ But UU' is symmetric matrix. Likewise, to verify (4) $XX^+ = S'\Lambda^{1/2}U'U\Lambda^{-1/2}S = S'S$. But $S'S$ is also symmetric.

2. Since $XX^-X = X$, from (2.1.5)
$$S'\Lambda^{1/2}U'X^-S'\Lambda^{1/2}U' = S'\Lambda^{1/2}U'. \tag{2.1.13a}$$

Premultiply (2.1.13a) by $\Lambda^{-1/2}S$ and postmultiply by $U\Lambda^{-1/2}$. The result then follows.

3. Since $UU' + VV' = I$ and $S'S + T'T = I$,
$$X^- = (UU' + VV')X^-(S'S + T'T)$$
$$= UU'X^-S'S + VV'X^-S'S + UU'X^-T'T + VV'X^-T'T$$
$$= U\Lambda^{-1/2}S + VV'X^-S'S + UU'X^-T'T + VV'X^-T'T$$
$$= X^+ + UA_1T + VA_2S + VA_3T$$
$$= [U \ V]\begin{bmatrix} \Lambda^{-1/2} & A_1 \\ A_2 & A_3 \end{bmatrix}\begin{bmatrix} S \\ T \end{bmatrix}. \tag{2.1.13b}$$

4. First
$$XMX = A_1A_2A_3 \tag{2.1.14a}$$

where

$$A_1 = [S' \; T'] \begin{bmatrix} \Lambda^{1/2} & 0 \\ 0 & 0 \end{bmatrix} \begin{bmatrix} U' \\ V' \end{bmatrix},$$

$$A_2 = [U \; V] \begin{bmatrix} \Lambda^{-1/2} & C_1 \\ C_2 & C_3 \end{bmatrix} \begin{bmatrix} S \\ T \end{bmatrix}$$

and

$$A_3 = [S' \; T'] \begin{bmatrix} \Lambda^{1/2} & 0 \\ 0 & 0 \end{bmatrix} \begin{bmatrix} U' \\ V' \end{bmatrix}.$$

Thus, from (2.1.14a)

$$XMX = A_1 A_2 A_3 = [S' \; T'] \begin{bmatrix} \Lambda^{1/2} & 0 \\ 0 & 0 \end{bmatrix} \begin{bmatrix} U' \\ V' \end{bmatrix} \tag{2.1.14b}$$
$$= X.$$

Notice from (2.1.12)

$$MX = [U \; V] \begin{bmatrix} I & 0 \\ C_2\Lambda^{1/2} & 0 \end{bmatrix} \begin{bmatrix} U' \\ V' \end{bmatrix}, \tag{2.1.15a}$$

and

$$XM = [S' \; T'] \begin{bmatrix} I & \Lambda^{1/2}C_1 \\ 0 & 0 \end{bmatrix} \begin{bmatrix} S \\ T \end{bmatrix}. \tag{2.1.15b}$$

From (2.1.15)

$$MXM = A_2 A_4 \tag{2.1.16a}$$

with

$$A_4 = [S' \; T'] \begin{bmatrix} I & \Lambda^{1/2}C_1 \\ 0 & 0 \end{bmatrix} \begin{bmatrix} S \\ T \end{bmatrix}.$$

Thus, from (2.1.16a)

$$MXM = [U \; V] \begin{bmatrix} \Lambda^{-1/2} & C_1 \\ C_2 & C_2\Lambda^{1/2}C_1 \end{bmatrix} \begin{bmatrix} S \\ T \end{bmatrix}. \tag{2.1.16b}$$

Thus from (2.1.15) $MXM = M$ iff $C_2\Lambda C_1 = C_3$. ∎

The following observations give a representation of $(X'X)^+$.

From (2.2.14), MX is symmetric iff $C_2 = 0$. From (2.1.15) XM is symmetric iff $C_1 = 0$.

Since

$$X = [S'\ T']\begin{bmatrix} \Lambda^{1/2} & 0 \\ 0 & 0 \end{bmatrix}\begin{bmatrix} U' \\ V' \end{bmatrix},$$

(2.1.17)

$$X'X = [U\ V]\begin{bmatrix} \Lambda & 0 \\ 0 & 0 \end{bmatrix}\begin{bmatrix} U' \\ V' \end{bmatrix}.$$

The following corollary of Theorem 2.1.2 results.

Corollary to Theorem 2.1.2.

1. $(X'X)^+ = U\Lambda^{-1}U'$.
2. $U'(X'X)^-U = \Lambda^{-1}$ for every generalized inverse of $X'X$.
3. A representation of $(X'X)^-$ is

$$(X'X)^- = [U\ V]\begin{bmatrix} \Lambda^{-1} & B_1 \\ B_1' & B_2 \end{bmatrix}\begin{bmatrix} U' \\ V' \end{bmatrix}$$

(2.1.18)

$$= (X'X)^+ + UB_1V' + VB_1'U' + VB_2V'$$

with $B_1 = U'(X'X)^-V$ and $B_2 = V'(X'X)^-V$.

4. The matrix

$$H = [U\ V]\begin{bmatrix} \Lambda^{-1} & B_1 \\ B_1' & B_2 \end{bmatrix}\begin{bmatrix} S \\ T \end{bmatrix}$$

is a generalized inverse of $X'X$. In Definition 2.1.3 conditions (3) and (4) are satisfied by $B_1 = 0$. Condition (2) is satisfied if $B_2 = B_1\Lambda B_1'$.

If N is non-singular matrix

$$(UNU')^+ = UN^{-1}U'.$$

(2.1.19)

This fact may be verified by straightforward verification of the defining properties of Moore-Penrose inverses.

Also let P be an orthogonal matrix. Then one may verify using the axioms

$$(PAP')^+ = PA^+P'. \tag{2.1.20}$$

Another result to be used in the sequel is

$$\begin{bmatrix} A^+ & 0 \\ 0 & B^+ \end{bmatrix} = \begin{bmatrix} A & 0 \\ 0 & B \end{bmatrix}^+. \tag{2.1.21}$$

The following examples illustrate some of the ideas above.

Example 2.1.8. An Ordinary Inverse Is a Generalized Inverse. Since X, $X'X$, and XX' in Example 2.1.6 are non-singular the generalized inverse is the ordinary inverse. Observe

$$X^{-1} = \frac{1}{8} \begin{bmatrix} \frac{2}{\sqrt{2}} & -\frac{2}{\sqrt{2}} \\ \frac{4}{\sqrt{2}} & \frac{4}{\sqrt{2}} \end{bmatrix},$$

$$(X'X)^{-1} = \begin{bmatrix} \frac{1}{16} & 0 \\ 0 & \frac{1}{4} \end{bmatrix},$$

and

$$(XX')^{-1} = \frac{1}{64} \begin{bmatrix} 10 & 6 \\ 6 & 10 \end{bmatrix}.$$

Example 2.1.9. Moore-Penrose Inverse. The Moore-Penrose

inverse of X in Example 2.1.4 is

$$X^+ = \begin{bmatrix} \frac{1}{\sqrt{2}} & \frac{1}{\sqrt{2}} & 0 \\ \frac{1}{\sqrt{2}} & -\frac{1}{\sqrt{2}} & 0 \\ 0 & 0 & 1 \end{bmatrix} \begin{bmatrix} \frac{1}{\sqrt{2}} & 0 & 0 \\ 0 & 1 & 0 \\ 0 & 0 & 0 \end{bmatrix} \begin{bmatrix} 1 & 0 & 0 \\ 0 & 1 & 0 \\ 0 & 0 & 1 \end{bmatrix}$$

$$= \begin{bmatrix} \frac{1}{2} & \frac{1}{\sqrt{2}} & 0 \\ \frac{1}{2} & -\frac{1}{\sqrt{2}} & 0 \\ 0 & 0 & 0 \end{bmatrix}.$$

Thus,

$$(X'X)^+ = \begin{bmatrix} \frac{1}{\sqrt{2}} & \frac{1}{\sqrt{2}} & 0 \\ \frac{1}{\sqrt{2}} & -\frac{1}{\sqrt{2}} & 0 \\ 0 & 0 & 1 \end{bmatrix} \begin{bmatrix} \frac{1}{2} & 0 & 0 \\ 0 & 1 & 0 \\ 0 & 0 & 0 \end{bmatrix} \begin{bmatrix} \frac{1}{\sqrt{2}} & \frac{1}{\sqrt{2}} & 0 \\ \frac{1}{\sqrt{2}} & -\frac{1}{\sqrt{2}} & 0 \\ 0 & 0 & 1 \end{bmatrix}$$

$$= \begin{bmatrix} \frac{3}{4} & -\frac{1}{2} & 0 \\ -\frac{1}{4} & \frac{3}{4} & 0 \\ 0 & 0 & 1 \end{bmatrix}$$

and

$$(XX')^+ = \begin{bmatrix} 1 & 0 & 0 \\ 0 & 1 & 0 \\ 0 & 0 & 1 \end{bmatrix} \begin{bmatrix} \frac{1}{2} & 0 & 0 \\ 0 & 1 & 0 \\ 0 & 0 & 0 \end{bmatrix} \begin{bmatrix} 1 & 0 & 0 \\ 0 & 1 & 0 \\ 0 & 0 & 1 \end{bmatrix}$$

$$= \begin{bmatrix} \frac{1}{2} & 0 & 0 \\ 0 & 1 & 0 \\ 0 & 0 & 0 \end{bmatrix}.$$

Example 2.1.10. Relationship between Moore-Penrose Inverses and Other Generalized Inverses. Let

$$X = \begin{bmatrix} 1_3 & 1_3 & 0 & 0 \\ 1_3 & 0 & 1_3 & 0 \\ 1_3 & 0 & 0 & 1_3 \end{bmatrix}.$$

Then

$$X'X = \begin{bmatrix} 9 & 3 \cdot 1_3' \\ 3 \cdot 1_3 & 3 \cdot 1_3 \end{bmatrix}.$$

A generalized inverse of $X'X$ is

$$G = \begin{bmatrix} 0 & 0 \\ 0 & \frac{1}{3} \cdot 1_3 \end{bmatrix}. \tag{2.1.22}$$

Now, an eigenvector of zero for $X'X$ is $V' = \frac{1}{2}[1 - 1 - 1 - 1]$. Thus,

$$VV' = \frac{1}{4}\begin{bmatrix} 1 & -1_3' \\ -1_3 & J_3 \end{bmatrix}$$

and

$$UU' = I - VV' = \begin{bmatrix} \frac{3}{4} & \frac{1}{4} \cdot 1_3' \\ \frac{1}{4} \cdot I_3 & I - \frac{1}{4} \cdot J_3 \end{bmatrix}.$$

Now

$$(X'X)^+ = UU'GUU' = \begin{bmatrix} \frac{1}{16} & \frac{1}{48} \cdot 1_3' \\ \frac{1}{48} \cdot 1_3 & \frac{1}{3}(I_3 - \frac{5}{16}J_3) \end{bmatrix}. \tag{2.1.23}$$

Exercise 2.1.6. Let

$$A = \begin{bmatrix} 4 & 4 & 1 \\ 4 & 4 & 1 \\ 1 & 1 & 4 \end{bmatrix}.$$

Verify that

$$G = \begin{bmatrix} 0 & 0 & 0 \\ 0 & \frac{4}{15} & -\frac{1}{15} \\ 0 & -\frac{1}{15} & \frac{4}{15} \end{bmatrix}$$

is a generalized inverse of A.

Exercise 2.1.7. Find the Moore-Penrose inverse of each matrix in Exercise 2.1.3. Also find another generalized inverse. Verify that $(X'X)^+ = UU'GUU'$ and $X' = UU'GSS'$.

Exercise 2.1.8. Repeat Exercise 2.1.7 for

$$X = \begin{bmatrix} 1 & 1 & 0 & 1 & 0 \\ 1 & 1 & 0 & 0 & 1 \\ 1 & 1 & 0 & 1 & 0 \\ 1 & 0 & 1 & 0 & 1 \\ 1 & 0 & 1 & 1 & 0 \\ 1 & 0 & 1 & 0 & 1 \end{bmatrix}.$$

Exercise 2.1.9. Show that $XX' = S\Lambda S'$ and that the Moore-Penrose inverse of $XX' = S\Lambda^{-1}S'$.

Exercise 2.1.10. Let A be a nonsingular n dimensional matrix. Show that $(S'AS)^+ = S'A^{-1}S$.

Exercise 2.1.11. For X in Exercises 2.1.3 and 2.1.8 find

$$(X'X + kI)^{-1}$$

with k a positive scalar.

Exercise 2.1.12. Show that $(X'X + VV')^{-1}$ is a generalized inverse of $X'X$.

The results on generalized inverses above are those that will be used in the sequel. More information about generalized inverses is available in sources like Rao and Mitra (1971), Graybill (1976) and Albert (1972).

2.2 The Bayes Estimator

The BE in the context of a linear model will be presented in Chapter III as the solution to an optimization problem. It will be presented here in a more general setting than in Chapter III. A clear indication of how Bayes Theorem is used in its derivation will be given. Let $Z = (Z_1, Z_2, \ldots, Z_m)$ be an m dimensional continuous random variable. Let $X = (X_1, X_2, \ldots, X_n)$ be an n dimensional random variable. The following notations will be employed for conditional density functions:

$$a(Z_1, Z_2, \cdots, Z_m | X_1, X_2, \ldots, X_n) = a(Z|X);$$

$$C(x) = C(x_1, x_2 \ldots x_n).$$

Furthermore define

$$\int a(z|x)C(x)dx = \int \cdot \cdot \int a(z|x)C(x)dx_1 dx_2 \ldots dx_n.$$

The prior distributions and the populations considered here will always be continuous.

Let (Θ, A, L) be a triple where Θ is the set of m dimensional parameters, A is a set of actions and L is a function whose domain is the Cartesian product set $\Theta \times A$ and whose range is the real numbers. Let X be a sample space of n dimensional vectors. Let $d(x)$ be a mapping from X to A. The triple (Θ, A, L) is called a decision space. The function L is called a loss function. The mapping $d(x)$ is the decision rule.

Let θ be the true value of the parameter. The risk function is defined by the equation

$$R(\theta, d) = E_\theta L(\theta, d(x)) = \int L(\theta, d(x)) f(x|\theta) \, dx. \qquad (2.2.1)$$

There are a number of different loss functions described in the literature. In this monograph the squared error loss function will be used; that is

$$L(\theta, d(x)) = (\theta - d(x))^2. \qquad (2.2.2)$$

For the squared error loss function in (2.2.2) the risk is the mean square error conditional on θ.

Suppose θ has a prior distribution $\pi(\theta)$. Let $X = (X_1, X_2 \ldots X_n)$ be a random sample from a population with parameter θ and probability density function $f(x|\theta)$. To find $g(\theta|x)$, the posterior distribution, Bayes theorem is used. Thus,

$$g(\theta|x) = \frac{f(x|\theta)\pi(\theta)}{\int f(x|\theta)\pi(\theta) \, d\theta}. \qquad (2.2.3)$$

The density $g(\theta|x)$ is called the posterior distribution.

A decision rule δ_0 is said to be Bayes with respect to a prior distribution $\pi(\theta)$ if

$$r(\pi, \delta_0) = \inf_{d \in D^*} r(\pi, d), \qquad (2.2.4)$$

where D^* is the set of all possible decision rules and

$$r(\pi, d) = ER(\theta, d) = \int_\Theta R(\theta, d)\pi(\theta) \, d\theta$$
$$= \int_\Theta \int_X (\theta - d(x))^2 f(x|\theta)\pi(\theta) \, dx \, d\theta. \qquad (2.2.5)$$

The quantity $r(\pi, d)$ is called the Bayes risk or the MSE averaging over the prior distribution.

Under certain conditions, that are assumed to be true here, by Fubini's theorem

$$r(\pi, d) = \int_X \int_\Theta (\theta - d(x))^2 g(\theta|x) \, d\theta \, dx. \qquad (2.2.6)$$

Thus the Bayes rule is the one that minimizes

$$L(x) = \int_{\Theta} (\theta - d(x))^2 g(\theta|x) \, d\theta. \qquad (2.2.7)$$

Differentiating under the integral sign with respect to $d(x)$, setting the derivative equal to zero and solving the resulting equation for $d(x)$ it is seen that the Bayes decision rule is

$$d(x) = \int_{\Theta} \theta g(\theta|x) \, d\theta, \qquad (2.2.8)$$

the posterior mean. Since $d(x)$ is the mean of the posterior distribution, the Bayes risk of $d(x)$ is the variance of that distribution.

The Bayes rule is a best estimator in the sense that:

1. It minimizes the average MSE.
2. It is an admissible estimator. That means that there is no other estimator with a uniformly smaller risk than the BE.

An estimator that frequently coincides with the BE is the minimax estimator presented in the next section.

An example of finding an explicit BE will now be given.

Example 2.2.1. An Explicit BE. Consider a random sample $X_1, X_2, \ldots, X_n$ from a normal population with mean θ and variance σ^2. Assume that θ has a normal prior distribution with mean θ_0 and variance σ_0^2. Then

$$\bar{X} \sim N\left(\theta, \frac{\sigma^2}{n}\right). \qquad (2.2.9)$$

The posterior distribution is found by substituting

$$f(\bar{x}|\theta) = \frac{1}{\sqrt{\frac{2\pi\sigma^2}{n}}} \exp\left[-\frac{(\bar{x} - \theta)^2}{2\sigma^2/n}\right] \qquad (2.2.10)$$

and

$$\pi(\theta) = \frac{1}{\sqrt{2\pi\sigma_0^2}} \exp\left[-\frac{(\theta - \theta_0)^2}{2\sigma_0^2}\right] \qquad (2.2.11)$$

into equation (2.2.3). After completing the square of θ and evaluation of the integral in the denominator (using properties of the normal distribution) it is found that

$$\theta | \bar{X} \sim N\left(\frac{\bar{X}\sigma_0^2 + \theta_0\sigma^2/n}{\sigma_0^2 + \sigma_0^2/n}, \frac{(\sigma^2/n)\sigma_0^2}{\sigma_0^2 + \sigma^2/n}\right). \tag{2.2.12}$$

Thus, the Bayes estimator is

$$\hat{\theta} = \frac{\sigma_0^2}{\sigma_0^2 + \sigma^2/n}\bar{X} + \frac{\sigma^2/n}{\sigma_0^2 + \sigma^2/n}\theta_0. \tag{2.2.13}$$

The factor in front of $\bar{X}$ and θ_0 in the expression for the BE $\hat{\theta}$ are the relative weights of sample and prior information. Notice that

$$\text{Var }(\bar{X}) = \sigma^2/n \tag{2.2.14}$$

while

$$\text{Var }(\hat{\theta}|\bar{X}) = \frac{(\sigma^2/n)\sigma_0^2}{\sigma_0^2 + \sigma^2/n} = \frac{\sigma^2}{n} - \frac{\sigma^4/n^2}{\sigma_0^2 + \sigma^2/n} \tag{2.2.15}$$

$$= \sigma_0^2 - \frac{\sigma_0^4}{\sigma_0^2 + \sigma^2/n}.$$

From (2.2.15) it is seen that

1. $\text{Var }(\hat{\theta}|\bar{X}) \le \text{Var }(\bar{X})$; (2.2.16)
   (The posterior variance is much smaller than that of the sample mean.)
2. $\text{Var }(\hat{\theta}|\bar{X}) \le \text{Var }(\theta)$. (2.2.17)
   (The variance of the posterior distribution is smaller than that of the prior distribution.)

Now the conditional expectation

$$E_\theta(\hat{\theta} - \theta)^2 = E_\theta(\hat{\theta} - E_\theta(\hat{\theta}))^2 + [E_\theta(\hat{\theta}) - \theta]^2$$

$$= \text{Var}(\hat{\theta}) + (\text{ BIAS } \hat{\theta})^2 \tag{2.2.18}$$

$$= \frac{\sigma^2\sigma_0^2}{n\sigma_0^2 + \sigma^2} + \frac{\sigma^4}{(n\sigma_0^2 + \sigma^2)^2}(\theta - \theta_0)^2.$$

Then

$$E_\theta(\hat\theta - \theta)^2 < \frac{\sigma^2}{n} \qquad (2.2.19)$$

if and only if

$$|(\theta - \theta_0)| < \frac{\sigma}{\sigma_0 \sqrt{n}}. \qquad (2.2.20)$$

Thus, the MSE of $\hat\theta$ conditional on θ is less than that of the ordinary sample mean $\bar{X}$ only for a range of values of θ.

The average MSE of a BE is always less than that of the usual maximum likelihood estimator (MLE). However, the conditional MSE of the BE is generally less than that of the MLE only for a range of values of θ. The problem of determining the range of values of the parameter where a BE has smaller MSE than LS will be taken up for the linear model in Chapter VII.

Exercise 2.2.1. Let $X_1, X_2, \ldots, X_n$ be a random sample from a binomial distribution $b(1, \theta)$. Assume that

$$\Pi(\theta) = \begin{cases} \frac{\Gamma(\gamma+\beta)}{\Gamma(\gamma)\Gamma(\beta)} \theta^{\gamma-1}(1-\theta)^{\beta-1}, & 0 < \theta < 1, \\ 0, & \text{elsewhere.} \end{cases}$$

The prior distribution of θ is a gamma distribution with parameters γ and β.

A. Find the posterior distribution of $\theta|Y$ where $Y = \sum_1^n X_i$.

B. Find the Bayes estimator of θ.

Exercise 2.2.2. Let $X_1, X_2, \ldots, X_n$ be independent random variables where $X_i \sim N(\theta_i, 1)$. Assume $\theta_i \sim N(0, A)$.

A. Show that the Bayes estimator is

$$\hat\theta_i = \left(1 - \frac{1}{A+1}\right) X_i.$$

B. Show that $\dfrac{p-2}{\sum_1^p X_i^2}$ is an unbiased estimator of $\dfrac{1}{A+1}$.

C. Show that if $p > 2$, then

$$\hat{\theta}_i = \left(1 - \frac{p-2}{\sum_{i=1}^p X_i^2}\right) X_i$$

has uniformly smaller risk and Bayes risk than X_i.

This is the celebrated James-Stein (1961) estimator. It shows that when $p > 2$ the maximum likelihood estimator X_i is inadmissible; it does not have the smallest risk of all the estimators of θ for every value of θ.

Exercise 2.2.3. Let X be a single observation from a population with distribution

$$f(x|\theta) = \theta e^{-\theta x} \quad x > 0.$$

Assume that θ has a prior distribution $\Pi(\theta) = e^{-\theta}, \theta > 0$.
A. Find the posterior distribution.
B. Find the BE for the squared error loss function in (2.2.1).
C. Show that in general for a loss function of the form

$$L(\theta, d(x)) = |\theta - d(x)| \quad \text{(absolute error loss)}$$

the BE is the median of the posterior distribution.
D. For the posterior distribution in A find the BE for the absolute error loss function in C above.

2.3 The Minimax Estimator

Let Ω be a parameter space. Let $\theta \in \Omega$. Let $d_0(x)$ be the decision function such that for all $\theta \in \Omega$

$$\max_{\theta \in \Omega} R[\theta, d_0(x)] \leq \max_{\theta \in \Omega} R[\theta, d(x)] \qquad (2.3.1)$$

for every decision function $d(x)$. Then $d_0(x)$ is called the minimax estimator.

Of special interest in this monograph is the case where θ is an m dimensional vector with components θ_i and A is an NND matrix. Let

$$\Omega = \{\theta : (\theta - \theta_0)'A(\theta - \theta_0) \le 1\}. \tag{2.3.2}$$

In the next chapter it will be seen how in a linear regression model the minimax estimator for a parameter space in the form (2.3.2) coincides with the BE. Also it will be shown that

$$m = \max_{\theta \in \Omega} R[\theta, d_0(x)] = r(\pi, d_0). \tag{2.3.3}$$

In other words, on the ellipsoid (2.3.2) the maximum MSE neglecting prior assumptions is for the BE equal to its average MSE or Bayes risk.

For further information about Bayes and minimax estimators an excellent source is Ferguson (1967).

Example 2.3.1. A Minimax Estimator. Let $X_1, X_2 \ldots X_n$ be a random sample from a normal population with mean θ and variance σ^2. The objective is to find the estimator of the form

$$\hat{\theta} = \theta_0 + c(\bar{X} - \theta_0) \tag{2.3.4}$$

where the maximum value of the MSE to be minimized when

$$|\theta - \theta_0| < \sigma_0. \tag{2.3.5}$$

The resulting estimator is the minimax estimator. Observe that

$$E(\hat{\theta} - \theta)^2 = E[c(\bar{X} - \theta_0) - (\theta - \theta_0)]^2$$

$$= c^2 E(\bar{X} - \theta_0)^2 - 2cE(\bar{X} - \theta_0)(\theta - \theta_0) + (\theta - \theta_0)^2. \tag{2.3.6}$$

Now $E(\bar{X}) = \theta$ and

$$E(\bar{X} - \theta_0)^2 = \frac{\sigma^2}{n} + (\theta - \theta_0)^2. \tag{2.3.7}$$

Thus, for the values of θ in (2.3.5)

$$E(\hat{\theta} - \theta)^2 = \frac{c^2\sigma^2}{n} + (1 - c)^2(\theta - \theta_0)^2$$

$$\leq \frac{c^2\sigma^2}{n} + (1 - c)^2\sigma_0^2.$$

(2.3.8)

Differentiating the right hand side of the inequality (2.3.8) with respect to c, equating the derivative to zero and solving the resulting equation yields

$$c = \frac{\sigma_0^2}{\sigma^2/n + \sigma_0^2}.$$

(2.3.9)

Thus,

$$\hat{\theta} = \theta_0 + \frac{\sigma_0^2}{\sigma^2/n + \sigma_0^2}(\bar{X} - \theta_0)$$

(2.3.10)

$$= \frac{\sigma_0^2}{\sigma_0^2 + \sigma^2/n}\bar{X} + \frac{\sigma^2/n}{\sigma_0^2 + \sigma^2/n}\theta_0.$$

This is the same estimator obtained in equation (2.2.13).

Also notice that for c in (2.3.9) the MSE in (2.3.8) becomes, after straightforward algebra,

$$E(\hat{\theta} - \theta)^2 = \frac{\sigma_0^2(\sigma^2/n)}{\sigma_0^2 + \sigma^2/n}.$$

(2.3.11)

This is the same as the variance of the posterior distribution in (2.2.15).

Exercise 2.3.1. Let $X_1, X_2 \ldots X_p$ be independent $N(\theta_i, 1)$ random variables. Show that the estimator that minimizes the maximum value of

$$m = \frac{1}{p}\sum_{i=1}^{p} E(\hat{\theta}_i - \theta_i)^2$$

on $\sum_{i=1}^{p} \theta_i^2 \leq A$ is that of Exercise 2.2.2 part A.

2.4 Criterion for Comparing Estimators: Theobald's 1974 Result

The criterion for comparison of estimators will usually be the mean square error (MSE). Let θ be an m dimensional vector of parameters and let $\hat{\theta}$ be an estimator of θ. With respect to a matrix loss function the MSE is the matrix

$$M = E(\hat{\theta} - \theta)(\hat{\theta} - \theta)'. \qquad (2.4.1)$$

Let A be an $m \times m$ non-negative definite matrix. The MSE with respect to the quadratic loss function is

$$m_A = E(\hat{\theta} - \theta)'A(\hat{\theta} - \theta). \qquad (2.4.2)$$

Theobald's (1974) result, Theorem 2.4.1 states that one estimator has a smaller MSE than another estimator for a matrix loss function iff it has a smaller MSE for an arbitrary quadratic loss function. The formal statement and proof follows.

Theorem 2.4.1. Theobald's Result (1974). *A symmetric $m \times m$ matrix A is NND iff* tr $AB \geq 0$ *for all non-negative definite B.*

Proof. Suppose tr $AB \geq 0$ for all NND B. Let $B = pp'$ where p is a vector. Then

$$\text{tr } App' = p'Ap \geq 0.$$

Thus, A is NND.

On the other hand suppose A is NND. The SVD of B is

$$B = P\Lambda P' = \sum_{i=1}^{n} \lambda_i P_i P_i', \qquad (2.4.3)$$

where $\lambda_1, \lambda_2 \ldots \lambda_n$ are the eigenvalues of B. The P_i are the columns of P and are orthogonal eigenvectors of B. Now

$$\text{tr}(AB) = \sum_{i=1}^{n} \lambda_i P_i' A P_i \geq 0 \qquad (2.4.4)$$

since A is non-negative definite and $\lambda_i \geq 0$. ∎

An important consequence of Theobald's result is:
Let $\hat{\theta}_1$ and $\hat{\theta}_2$ be estimators of a parameter θ. (θ_1, θ_2 and θ are m dimensional column vectors.) Then

$$E(\hat{\theta}_1 - \theta)(\hat{\theta}_1 - \theta)' \leq E(\hat{\theta}_2 - \theta)(\hat{\theta}_2 - \theta)' \qquad (2.4.5)$$

iff for every NND matrix A,

$$E(\hat{\theta}_1 - \theta)'A(\hat{\theta}_1 - \theta) \leq E(\hat{\theta}_2 - \theta)'A(\hat{\theta}_2 - \theta). \qquad (2.4.6)$$

The matrix
$$M_i = E(\hat{\theta}_i - \theta)(\hat{\theta}_i - \theta)' \qquad (2.4.7)$$

is called the mean dispersion error (MDE) or mean square error (MSE) with respect to a matrix loss function. The quantity

$$m_i = E(\hat{\theta}_i - \theta)'A(\hat{\theta}_i - \theta) \qquad (2.4.8)$$

is the MSE averaging over a quadratic loss function.

Thus it has been established that estimator $\hat{\theta}_1$ has a smaller MDE than $\hat{\theta}_2$ iff the MSE of $\hat{\theta}_1$ averaging over every quadratic loss function is less than that of $\hat{\theta}_2$.

From the definition of positive definite matrices this means that $p'\hat{\theta}_1$ has a smaller MSE than $p'\hat{\theta}_2$ for all parametric functions if and only if $\hat{\theta}_1$ has a smaller MSE than $\hat{\theta}_2$ for all quadratic loss functions. Thus if $\hat{\theta}_1$ has a smaller MSE than $\hat{\theta}_2$ for all rank one quadratic loss functions the same result follows for all NND quadratic loss functions.

Let θ_i have components θ_{ij} $1 \leq j \leq m$. Let D be a diagonal matrix with elements d_j. The individual MSE of θ_{ij} is simply

$$m_{ij} = E(\hat{\theta}_{ij} - \theta_{ij})^2. \qquad (2.4.9)$$

The following simple theorem will be of use in Chapter VII.

Theorem 2.4.2. *Let D be any diagonal matrix. Then*

$$m_{1j} = E(\hat{\theta}_{1j} - \theta_{1j})^2 \leq E(\hat{\theta}_{2j} - \theta_{2j})^2 = m_{2j} \qquad (2.4.10)$$

iff

$$E(\hat{\theta}_1 - \theta_1)'D(\hat{\theta}_1 - \theta_1) \le E(\hat{\theta}_2 - \theta_2)'D(\hat{\theta}_2 - \theta_2). \qquad (2.4.11)$$

Proof. To see that (2.4.11) is an easy consequence of (2.4.10) notice that

$$E(\hat{\theta}_i - \theta_i)'D(\hat{\theta}_i - \theta_i) = \sum_{j=1}^{m} d_j E(\hat{\theta}_{ij} - \theta_{ij})^2 = \sum_{j=1}^{m} d_i m_{ij}. \qquad (2.4.12)$$

If (2.4.11) holds for all D, (2.4.10) is just the special case where $d_j = 1$ and $d_k = 0$ when $k \ne j$. ∎

2.5 Some Useful Inequalities

The following inequalities will be useful in the comparison of the MSE of estimators. The first one is very well known.

Theorem 2.5.1. Cauchy-Schwarz Inequality (CS). *Let b and c be nonzero $n \times 1$ column matrices. Then*

$$[b'c]^2 \le (b'b)(c'c). \qquad (2.5.1)$$

Equality holds only when b and c are scalar multiples of each other.

Proof. Let λ be a scalar. Then

$$0 \le (c - \lambda b)'(c - \lambda b) = c'c - 2\lambda b'c + \lambda^2 b'b. \qquad (2.5.2)$$

Inequality (2.5.2) holds iff

$$4(b'c)^2 - 4(b'b)(c'c) \le 0. \qquad (2.5.3)$$

Inequality (2.5.1) follows from (2.5.3). When equality holds in (2.5.1) it holds in (2.5.3). This implies equality in (2.5.2). Thus $c = \lambda b$. ∎

Corollary 2.5.1. *Let b and c be as in the CS inequality above. Let A be an $n \times n$ PD matrix. Then*

$$[b'c]^2 \leq b'A^{-1}bc'Ac. \tag{2.5.4}$$

Proof. By the CS inequality

$$[b'c]^2 = [b'A^{-1/2}A^{1/2}c]^2 \leq b'A^{-1}bc'Ac. \ \blacksquare \tag{2.5.5}$$

The following result is due to Farebrother (1976).

Theorem 2.5.2. *Let d be a positive scalar. A a PD matrix. Then $dA - bb'$ is positive semidefinite iff $b'A^{-1}b \leq d$.*

Proof. Notice that $dA - b'b$ is PD iff, for all $c \neq 0$,

$$dc'Ac > c'bbc = (c'b)^2. \tag{2.5.6}$$

From (2.5.6)

$$d > \frac{(c'b)^2}{c'Ac}. \tag{2.5.7}$$

Thus, for all $c \neq 0$,

$$\frac{d}{b'A^{-1}b} > \frac{(c'b)^2}{c'Acb'A^{-1}b}. \tag{2.5.8}$$

Then by the CS inequality since the maximum value of the right hand side of (2.5.7) is 1 and (2.5.7) holds for all c,

$$\frac{d}{b'A^{-1}b} \geq 1 \quad \text{or} \quad b'A^{-1}b \leq d. \ \blacksquare$$

The following result is a consequence of Theorem 2.5.2. It will be used to compare the MSE of the BE in 7.5.

Theorem 2.5.3. *Let A be a symmetric $n \times n$ matrix, a an $n \times 1$ vector and d a positive scalar. Then $dA - aa'$ is positive semidefinite iff A is non-negative definite, a belongs to the range of A, and $a'A^+a \leq d$.*

Proof. The singular value decomposition of A is

$$A = [P \ Q] \begin{bmatrix} \Lambda & 0 \\ 0 & 0 \end{bmatrix} \begin{bmatrix} P' \\ Q' \end{bmatrix} = P\Lambda P'. \qquad (2.5.9)$$

Then $dA - aa'$ is positive semidefinite iff

$$M = d \begin{bmatrix} \Lambda & 0 \\ 0 & 0 \end{bmatrix} - \begin{bmatrix} P'aa'P & P'aa'Q \\ Q'aa'P & Q'aa'Q \end{bmatrix} \qquad (2.5.10)$$

is positive semidefinite. But M is positive semidefinite iff $d\Lambda - P'aa'P$ is positive semidefinite, and $Q'aa'Q = 0$. But $Q'aa'Q = 0$ iff $Q'a = 0$ or a belongs to the range of A. The result then follows from $(2.5.10)$ since M is positive semidefinite iff PMP' is positive semidefinite.

The following result of Teräsvirta (1980) will be used to compare estimators.

Theorem 2.5.4. *Let b and c be two linearly independent $m \times 1$ vectors. Then $bb' - cc'$ is indefinite. Its nonzero eigenvalues are*

$$\lambda = \frac{-(b'b - c'c) \pm [(b'b - c'c)^2 + 4b'bc'c - 4(b'c)^2]^{1/2}}{2}. \qquad (2.5.11)$$

Proof. It must be shown that the eigenvalues have different signs. Let λ be an eigenvalue of $bb' - cc'$. Let z be an eigenvector. Then

$$(cc' - bb')z = \lambda z. \qquad (2.5.12)$$

Multiply $(2.5.12)$ on the left by b' and obtain

$$(b'cc' - b'bb')z = \lambda b'z \qquad (2.5.13)$$

or

$$(b'b + \lambda)b'z = b'cc'z. \qquad (2.5.14)$$

Multiply $(2.5.12)$ on the left by c'. Then

$$(c'c - \lambda)c'z = c'bb'z. \qquad (2.5.15)$$

From (2.5.14)

$$c'z = (b'c)^{-1}(b'b + \lambda)b'z. \tag{2.5.16}$$

Substitute (2.5.16) in (2.5.15). Then

$$(c'c - \lambda)(b'c)^{-1}(b'b + \lambda)b'z = c'bb'z \tag{2.5.17}$$

and since $(b'c)$ and $b'z$ are scalars,

$$(c'c - \lambda)(b'b + \lambda) = (c'b)^2. \tag{2.5.18}$$

Solve (2.5.18) to obtain the roots λ as in (2.5.11). But by the CS inequality

$$(b'c)^2 \leq b'bc'c \tag{2.5.19}$$

and thus the roots have different signs. ∎

Example 2.5.1. Illustration of Theorem 2.5.3. Let

$$b = \begin{bmatrix} 9 \\ 3 \\ 5 \end{bmatrix}, \quad c = \begin{bmatrix} 4 \\ -2 \\ 1 \end{bmatrix}.$$

It will be shown that $bb' - cc'$ is not positive semidefinite. Now the matrix

$$bb' - cc' = \begin{bmatrix} 81 & 27 & 45 \\ 27 & 9 & 15 \\ 45 & 15 & 25 \end{bmatrix} - \begin{bmatrix} 16 & -8 & 4 \\ -8 & 4 & -2 \\ 4 & -2 & 1 \end{bmatrix}$$

$$= \begin{bmatrix} 65 & 35 & 41 \\ 35 & 5 & 17 \\ 41 & 17 & 24 \end{bmatrix}$$

has rank 2. The characteristic equation is

$$\begin{vmatrix} 65 - \lambda & 35 & 41 \\ 35 & 5 - \lambda & 17 \\ 41 & 17 & 24 - \lambda \end{vmatrix} = 0$$

or

$$\lambda^3 - 94\lambda^2 - 1021\lambda = 0.$$

Thus, the eigenvalues are 0, 103.8 and -9.835.

Exercise 2.5.1. Show that
A. The matrix bb' is of rank 1 with non-zero eigenvalue $\sum_{i=1}^n b_i^2$.
B. The matrix $bb' - cc'$ has rank 2 for any dimension of vector b and c. (Assume b and c have same dimension.)

Exercise 2.5.2. Let

$$b = \begin{bmatrix} 4 \\ 2 \\ 1 \end{bmatrix}, \quad c = \begin{bmatrix} 1 \\ 2 \\ 2 \end{bmatrix}.$$

Find the eigenvalues of $bb' - cc'$.

Exercise 2.5.3. Find the center and axis of the ellipse

$$[\beta_1 - 1, \beta_2 + 2] \begin{bmatrix} \beta_1 - 1 \\ \beta_2 + 2 \end{bmatrix} \leq \begin{bmatrix} \frac{3}{8} & -\frac{1}{8} \\ -\frac{1}{8} & \frac{3}{8} \end{bmatrix}.$$

Exercise 2.5.4. Suppose b and c are linearly dependent. Under what conditions is $bb' - cc'$ positive definite (if any)? Illustrate with an example.

2.6 Some Miscellaneous Useful Matrix Results

In this section some matrix theory concepts and results will be presented that are useful in the study of Analysis of Variance. Their main use will be in Chapter X.

2.6.1 The Kronecker Product

Let $A = (a_{ij})$ and $B = (b_{ij})$ be $m \times n$ and $p \times q$ matrices respectively. The Kronecker product (also called the direct product)

$$A \otimes B = (a_{ij}B)$$

is an $mp \times nq$ matrix expressible as a partitioned matrix with $a_{ij}B$ as the $(ij)^{\text{th}}$ partition.

To illustrate a simple example is now given. Let

$$J_{2 \times 2} = \begin{bmatrix} 1 & 1 \\ 1 & 1 \end{bmatrix}$$

and

$$I_2 = \begin{bmatrix} 1 & 0 \\ 0 & 1 \end{bmatrix}.$$

Then

$$J \otimes I = \begin{bmatrix} 1 & 0 & 1 & 0 \\ 0 & 1 & 0 & 1 \\ 1 & 0 & 1 & 0 \\ 0 & 1 & 0 & 1 \end{bmatrix}$$

and

$$I \otimes J = \begin{bmatrix} 1 & 1 & 0 & 0 \\ 1 & 1 & 0 & 0 \\ 0 & 0 & 1 & 1 \\ 0 & 0 & 1 & 1 \end{bmatrix}.$$

Properties of the Kronecker product are developed by C.R. Rao (1973). The main application here will be to the Analysis of Variance. For this purpose the properties stated below are the most useful.

(i) $A_1 A_2 \otimes B_1 B_2 = (A_1 \otimes B_1)(A_2 \otimes B_2)$;
(ii) $(A \otimes B)^{-1} = A^{-1} \otimes B^{-1}$; (2.6.1)
(iii) $(A \otimes B)^- = A^- \otimes B^-$ for all G inverses.

Exercise 2.6.1. Let

$$A = \begin{bmatrix} 1 & 2 \\ 6 & 4 \end{bmatrix}, \quad B = \begin{bmatrix} -1 & 2 & 4 \\ 4 & -6 & 1 \end{bmatrix}.$$

Find $A \otimes B$.

Exercise 2.6.2. Find a generalized inverse of $X'X$ given

$$X = [1_2 \otimes 1_3 \ I_2 \otimes 1_3 \ 1_2 \otimes I_3].$$

Exercise 2.6.3. Show that $(A \otimes B)^+ = A^+ \otimes B^+$.

2.6.2 Patterned Matrices

The results on inverses of patterned matrices stated below will be used. The proof can be found in Graybill (1969).

Theorem 2.6.1. *Suppose an* $(m+n) \times (m+n)$ *matrix* C *is defined by*

$$C = \begin{bmatrix} a_1 I_m & a_2 J_{m \times n} \\ a_2 J_{m \times n} & a_3 I_n \end{bmatrix} \tag{2.6.2}$$

where $a_3 \neq 0$, $n > 0$. *Then*

(1) The inverse of C *exists iff*

$$a_1 \neq 0, \quad a_1 \neq \frac{mna_2^2}{a_3}.$$

(2) If C^{-1} *exists*

$$C^{-1} = \begin{bmatrix} \frac{1}{a_1} I_m + b_1 J_{m \times m} & b_2 J_{m \times m} \\ b_2 J'_{m \times m} & \frac{1}{a_3} I_m + b_3 J_{n \times n} \end{bmatrix} \tag{2.6.3}$$

where

$$b_1 = -\frac{na_2}{a_1(mna_2^2 - a_3a_1)},$$

$$b_2 = \frac{a_2}{mna_2^2 - a_3a_1}$$

and

$$b_3 = -\frac{ma_2^2}{a_3(mna_2^2 - a_3a_1)}.$$

Exercise 2.6.4. Find the inverse of

$$A = \begin{bmatrix} 2 & 0 & 1 & 1 \\ 0 & 2 & 1 & 1 \\ 1 & 1 & 3 & 0 \\ 1 & 1 & 0 & 3 \end{bmatrix}.$$

2.7 Summary

The following useful topics from matrix theory were presented:

1. positive semidefinite matrices;
2. the singular value decomposition;
3. generalized inverses;
4. the Kronecker Product;
5. patterned matrices.

The results are particularly useful in matrix representation of estimators and their efficiencies.

Bayes and Minimax estimators were presented. Their equivalence in a regression context will be established in Chapter IV.

The inequalities presented in Section 2.5 will allow for comparison of different BE and comparison of BE and LS estimators.

PART II
THE ESTIMATORS

Chapter III

The Estimators

3.0 Introduction

Five kinds of estimators will be presented in this chapter. These include

1. least square estimators;
2. the generalized ridge regression estimator;
3. the mixed estimators;
4. the minimax estimator;
5. the linear Bayes estimator.

To obtain each kind of estimator a different optimization problem will be solved. Thus, each estimator is optimum with respect to a certain criterion of goodness. It should be observed that some of the optimization problems have some common elements.

In Section 3.1 the concept of estimability is defined and seen to be the criterion for uniqueness of least square estimators for the non-full rank model. The proof of the Gauss-Markov Theorem is given. This theorem states that the least square estimator is the unbiased linear estimator with the smallest variance.

The generalized ridge regression estimator is derived in Section 3.2. The optimization problem that is solved there is a generalization of the one used by Hoerl and Kennard (1970) to obtain the ordinary ridge estimator presented in Section 1.3.

The mixed estimators that are derived in Section 3.3 are really a special kind of least square estimator. The experimenter takes some additional observations and puts together an augmented linear model and then obtains the least square estimator for that model.

The presentation of the linear minimax estimators in Section 3.4 is a modification of that found in Toutenburg (1982), Section 4.2, beginning on page 84. However, Toutenburg also presents a minimax estimator for a general quadratic loss function which depends on the loss function matrix. By Theobald's result derived in Section 2.4 the minimax estimator presented in Section 3.4 should also be minimax for an arbitrary quadratic loss function.

The derivation of the linear Bayes estimator follows that of C.R. Rao (1973). When the prior mean is unknown and the parametric functions are estimable the optimum estimator is the least square estimator. An alternative derivation of the linear Bayes estimator in terms of the least square estimator is also given. In Chapter IV the two forms of the estimator will be shown to be equivalent.

3.1 The Least Square Estimator and Its Properties

Consider the linear model

$$Y = X\beta + \epsilon, \tag{3.1.1a}$$

where X has rank $s \leq m$.

In Section 1.2 the least square estimator was obtained by minimizing the sum of the squares of the deviations between the observed and the predicted values. The form of the least square estimator was given for the full and the non-full rank case. The reader will recall from Chapter II that a generalized inverse need not be unique. Thus, it is certainly possible that the parameters have more than one least square estimator. Let G_1 and G_2 be two distinct generalized inverses of $X'X$. Then two expressions for the least square estimator of $p'\beta$ are

$$p'b_1 = p'G_1 X'Y \tag{3.1.1b}$$

and

$$p'b_2 = p'G_2X'Y. \tag{3.1.1c}$$

The natural question to ask is, for what class of parametric functions is $p'b$ unique?

Notice that from (2.1.18) any LS estimator may be written in the form

$$b = (X'X)^+X'Y + UU'GVV'X'Y$$

$$+ VV'GUU'X'Y + VV'GVV'X'Y \tag{3.1.2}$$

$$= (X'X)^+X'Y + VV'GX'Y$$

from (2.1.7) and (2.1.8).

Thus, from (3.1.2), $p'b$ is unique for all generalized inverses if

$$p'V = 0. \tag{3.1.3}$$

Parametric functions that satisfy (3.1.3) are called estimable functions. Farebrother (1976, 1978) defines estimability by stating that $p'\beta$ is estimable iff

$$p = Ud \quad \text{for some } s \text{ dimensional vector } d. \tag{3.1.4}$$

This condition for estimability will be used extensively in the sequel.

Many authors (see for example Searle (1971)) define estimability by saying that $p'\beta$ is the expectation of a linear function of the observations. In other words, there is a linear function of the observations that is an unbiased estimator of $p'\beta$. Thus,

$$p'\beta = E(c'Y) = c'X\beta \tag{3.1.5}$$

or equivalently,

$$p' = c'X. \tag{3.1.6}$$

To see that (3.1.3), (3.1.4) and (3.1.6) are equivalent it will be shown that (3.1.3) $\rightarrow$ (3.1.4) $\rightarrow$ (3.1.6) $\rightarrow$ (3.1.3).

For the first implication, $(3.1.3) \rightarrow (3.1.4)$, observe that since $[U \ V]$ is an orthogonal matrix $p' = p'(UU' + VV') = pUU'$. The required $d = U'p$.

To see that $(3.1.4) \rightarrow (3.1.6)$, observe that since $X' = U\Lambda^{1/2}S$, $U = X'S'\Lambda^{-1/2}$. Then $p = X'S^{-1/2}d$ is all that is needed.

To show $(3.1.6) \rightarrow (3.1.3)$ notice that, since $p' = c'X$ and $XV = 0$, $p'V = c'XV = 0$.

When X is of full rank, $(3.1.6)$ has a unique solution. Thus, for the full rank case, every parametric function is estimable.

Example 3.1.1. Estimable and Non-Estimable Parametric Functions. Consider the linear model

$$
\begin{bmatrix} Y_1 \\ Y_2 \\ Y_3 \end{bmatrix} = \begin{bmatrix} 1 & 1 & 0 \\ \frac{1}{\sqrt{2}} & \frac{1}{\sqrt{2}} & 0 \\ 0 & 0 & 1 \end{bmatrix} \begin{bmatrix} \beta_0 \\ \beta_1 \\ \beta_2 \end{bmatrix} + \begin{bmatrix} \epsilon_1 \\ \epsilon_2 \\ \epsilon_3 \end{bmatrix}.
$$

An LS estimator is

$$
\begin{bmatrix} \hat{\beta}_0 \\ \hat{\beta}_1 \\ \hat{\beta}_2 \end{bmatrix} = \begin{bmatrix} 0 \\ \frac{2}{3}\left(Y_1 + \frac{Y_2}{\sqrt{2}}\right) \\ Y_3 \end{bmatrix}.
$$

Notice that the SVD of $X'X$ is

$$X'X = \begin{bmatrix} \frac{1}{\sqrt{2}} & 0 & \frac{1}{\sqrt{2}} \\ \frac{1}{\sqrt{2}} & 0 & -\frac{1}{\sqrt{2}} \\ 0 & 1 & 0 \end{bmatrix} \begin{bmatrix} 3 & 0 & 0 \\ 0 & 1 & 0 \\ 0 & 0 & 0 \end{bmatrix} \begin{bmatrix} \frac{1}{\sqrt{2}} & \frac{1}{\sqrt{2}} & 0 \\ 0 & 0 & 1 \\ \frac{1}{\sqrt{2}} & -\frac{1}{\sqrt{2}} & 0 \end{bmatrix}$$

$$= \begin{bmatrix} \frac{1}{\sqrt{2}} & 0 \\ \frac{1}{\sqrt{2}} & 0 \\ 0 & 1 \end{bmatrix} \begin{bmatrix} 3 & 0 \\ 0 & 1 \end{bmatrix} \begin{bmatrix} \frac{1}{\sqrt{2}} & \frac{1}{\sqrt{2}} & 0 \\ 0 & 0 & 1 \end{bmatrix} .$$

Thus, $U = \begin{bmatrix} \frac{1}{\sqrt{2}} & 0 \\ \frac{1}{\sqrt{2}} & 0 \\ 0 & 1 \end{bmatrix}$ and $p'\beta$ is estimable iff there is a vector $\begin{bmatrix} d_1 \\ d_2 \end{bmatrix}$

such that

$$\begin{bmatrix} p_0 \\ p_1 \\ p_2 \end{bmatrix} = \begin{bmatrix} \frac{1}{\sqrt{2}} & 0 \\ \frac{1}{\sqrt{2}} & 0 \\ 0 & 1 \end{bmatrix} \begin{bmatrix} d_1 \\ d_2 \end{bmatrix} .$$

Thus $\beta_0 + \beta_1 + \sqrt{2}\beta_2$ is estimable since $d_1 = d_2 = \sqrt{2}$. But $\beta_0 - \beta_1$

is not estimable since the system

$$1 = \frac{1}{\sqrt{2}}d_1 \; ,$$

$$-1 = \frac{1}{\sqrt{2}}d_2 \; ,$$

$$0 = d_2$$

has no solution.

Example 3.1.2. Uniqueness of Estimable Parametric Functions. Consider the linear model

$$Y = \begin{bmatrix} 1_3 & 1_3 & 0 & 0 \\ 1_3 & 0 & 1_3 & 0 \\ 1_3 & 0 & 0 & 1_3 \end{bmatrix} \begin{bmatrix} \mu \\ \alpha_1 \\ \alpha_2 \\ \alpha_3 \end{bmatrix} + \epsilon.$$

From Example 2.1.8 it is seen that the normal equations are

$$9\hat{\mu} + 3\sum_{i=1}^{3} \hat{\alpha}_i = Y_{..},$$

$$3\hat{\mu} + 3\hat{\alpha}_i = Y_{i.}, \quad 1 \leq i \leq 3,$$

where $Y_{..} = \sum_{i=1}^{3} \sum_{j=1}^{3} Y_{ij}$ and $Y_{i.} = \sum_{j=1}^{3} Y_{ij}$. Using the generalized inverses in (2.1.24) and (2.1.25), two solutions to these normal equations are

$$\hat{u} = 0$$

$$\hat{\alpha}_i = \frac{Y_{i.}}{3}, \quad 1 \le i \le 3$$

and

$$\hat{\mu} = \frac{1}{12}Y_{..}$$

$$\hat{\alpha}_i = \frac{1}{48}Y_{..} + \frac{1}{3}Y_{i.} - \frac{5}{48}Y_{..}$$

$$= \frac{1}{3}Y_{i.} - \frac{1}{12}Y_{..}, \quad 1 \le i \le 3.$$

The estimator $\hat{\alpha}_2$ is not estimable. The equations $p' = t'X$ are

$$0 = \sum_{i=1}^{9} t_i,$$

$$0 = \sum_{i=1}^{3} t_i,$$

$$1 = \sum_{i=4}^{6} t_i,$$

$$0 = \sum_{i=7}^{9} t_i.$$

These equations have no solution. This can be seen by comparing the first equation with the sum of the other three. On the other hand $\hat{\alpha}_1 + \hat{\alpha}_2 - 2\hat{\alpha}_3$ is estimable. The solution to the system

$$0 = \sum_{i=1}^{9} t_i,$$

$$1 = \sum_{i=1}^{3} t_i,$$

$$1 = \sum_{i=4}^{6} t_i,$$

$$-2 = \sum_{i=7}^{9} t_i$$

is $t_1 = t_2 = t_3 = t_4 = t_5 = t_6 = 1/3$ and $t_7 = t_8 = t_9 = -2/3$.

Notice that two different estimators of α_2 were obtained but for basic solutions to the normal equations

$$\hat{\alpha}_1 + \hat{\alpha}_2 - 2\hat{\alpha}_3 = \frac{1}{3}(Y_{1.} + Y_{2.} - 2Y_{3.}).$$

Exercise 3.1.1.   A. Find two LS estimators for β when

$$\begin{bmatrix} Y_{11} \\ Y_{12} \\ Y_{21} \\ Y_{22} \\ Y_{31} \\ Y_{32} \end{bmatrix} = \begin{bmatrix} 1_2 & 1_2 & 0 & 0 \\ 1_2 & 0 & 1_2 & 0 \\ 1_2 & 0 & 0 & 1_2 \end{bmatrix} \begin{bmatrix} \mu \\ \alpha_1 \\ \alpha_2 \\ \alpha_3 \end{bmatrix} + \epsilon.$$

B. Show that $\mu + \alpha_i$, $1 \leq i \leq 3$, are estimable parametric functions and linearly independent.

C. Show that all estimable parametric functions may be expressed as a linear combination of the parametric functions in A.

D. Show that the estimators of $\alpha_1 - \alpha_2$ and $\alpha_1 + \alpha_2 - 2\alpha_3$ are the same for the LS estimators in A. However the estimators of $\alpha_1 + \alpha_2 + \alpha_3$ are not the same. How do you account for this?

Recall that $\hat{\theta}$ is an unbiased estimator of θ iff

$$E(\hat{\theta}) = \theta. \tag{3.1.7}$$

Estimators of the form $L'Y$ are called linear estimators. The Gauss-Markov theorem says that the LS estimator of $p'\beta$ is the best linear unbiased estimator. Of all the linear unbiased estimators the LS estimator has the smallest variance. More formally:

Theorem (Gauss-Markov) 3.1.1. *Let*

$$Y = X\beta + \epsilon \tag{3.1.8a}$$

be the classical linear regression model of full or non-full rank with

$$E(\epsilon) = 0 \quad \text{and} \quad D(\epsilon) = \sigma^2 I. \tag{3.1.8b}$$

Then: (1) The linear estimator $L'Y$ is unbiased for the parametric function $p'\beta$ iff $p'\beta$ is estimable. (2) The LS estimator is the minimum variance unbiased linear estimator (MVUE) for the estimable parametric functions.

Proof. (a) To prove (1), observe that

$$E[L'Y] = L'X\beta = p'\beta \tag{3.1.9}$$

holds true iff $p' = L'X$ or $p'\beta$ is estimable. (b) The variance of $L'Y$ is

$$V(L'Y) = L'L\sigma^2. \tag{3.1.10}$$

The theorem will be proved if it can be shown that when (3.1.10) is minimized subject to the constraint $p' = L'X$ (the estimability condition) the optimum estimator is the LS.

Let λ be a column vector of Lagrange multipliers. Consider

$$H = L'L\sigma^2 + (L'X - p')\lambda. \tag{3.1.11}$$

Differentiating H with respect to L, setting the result equal to zero yields

$$L\sigma^2 + X\lambda = 0. \tag{3.1.12}$$

The estimability condition with matrices transposed is

$$X'L - p = 0. \tag{3.1.13}$$

Equations (3.1.12) and (3.1.13) will now be solved simultaneously for λ and L. Multiply (3.1.12) by X'. Thus,

$$X'L\sigma^2 + X'X\lambda = 0. \tag{3.1.14}$$

Let G be a generalized inverse of $X'X$. Then from (3.1.12) and (3.1.13)

$$\lambda = -GX'L\sigma^2 = -Gp\sigma^2. \tag{3.1.15}$$

Substituting the expression for λ in (3.1.15) in (3.1.12), the optimum

$$L = XGp. \tag{3.1.16}$$

Thus, the minimum variance unbiased linear estimator is

$$p'\hat{\beta} = p'G'X'Y = p'(X'X)^+X'Y \tag{3.1.17}$$

since $p'\beta$ is estimable. (If G is a generalized inverse of $X'X$ so is G'.) Equation (3.1.17) is the LS estimator. ∎

Exercise 3.1.2. A. Consider a random sample of size three from a population with mean μ and variance σ^2. Show that the MVUE is the sample mean. The geometric analogue of this problem is that of finding the point in the plane $X + Y + Z = 1$ closest to the origin.

B. Repeat part A for a random sample of size n.

The geometric analogue of the Gauss-Markov Theorem is that of finding the point in n dimensional space on the hyperplanes $L'X = p'$ that lies closest to the origin.

Sometimes there may be correlations between the observations or some of the observations should receive more weight than others. An optimum estimator for this situation is the weighted least square estimator. To obtain it, minimize the weighted sum of squares of the differences between the observed and the predicted values. Thus, minimize

$$H = (Y - X\beta)'V^{-1}(Y - X\beta) \tag{3.1.18}$$

to obtain

$$p'b = p'(X'V^{-1}X')^{+}X'V^{-1}Y. \tag{3.1.19}$$

The Gauss-Markov theorem holds true if (3.1.8) is replaced by

$$E(\epsilon) = 0 \quad \text{and} \quad D(\epsilon) = V. \tag{3.1.20}$$

The problem of finding the BLUE when V is singular is considered in C.R. Rao (1973).

Exercise 3.1.3. Consider the model

$$\begin{bmatrix} Y_{11} \\ Y_{12} \\ Y_{21} \\ Y_{22} \end{bmatrix} = \begin{bmatrix} 1 & 0 \\ 1 & 0 \\ 0 & 1 \\ 0 & 1 \end{bmatrix} \begin{bmatrix} \alpha_1 \\ \alpha_2 \end{bmatrix} + \epsilon,$$

where the error vector satisfies the conditions

$$E(\epsilon) = 0 \quad \text{and} \quad D(\epsilon) = \frac{\sigma^2}{4}(4I - J) = V.$$

Show that α is a solution to

$$X'V^- X\alpha = X'V^- Y$$

(X is the design matrix) iff $\hat{\alpha}_1 - \hat{\alpha}_2 = \bar{Y}_1. - \bar{Y}_2..$

3.2 The Generalized Ridge Regression Estimator

The ridge regression estimator was obtained by minimizing the distance from the vector B to the origin, subject to the side condition that B lay on the ellipsoid

$$(B - b_1)'X'X(B - b_1) = \phi_0. \tag{3.2.1}$$

Perhaps on the basis of some prior information it might be better to minimize the distance from a point other than the origin. Also, perhaps some variables should receive more weight than others. Thus, instead of minimizing $B'B$, minimize the weighted distance

$$D = (B - \theta)'H(B - \theta) \tag{3.2.2}$$

subject to (3.2.1) where H is a positive semidefinite matrix.
   Let λ be the scalar Lagrange multiplier. Differentiate

$$L = (B - \theta)'H(B - \theta) + \lambda[(B - b_1)'X'X(B - b_1) - \phi_0] \tag{3.2.3}$$

to obtain

$$H(B - \theta) + \lambda X'X(B - b_1) = 0. \tag{3.2.4}$$

Thus the optimum estimator is

$$\hat{\beta}gr = [H + \lambda X'X]^+[H\theta + \lambda X'Xb_1], \tag{3.2.5}$$

the generalized ridge regression estimator.
   When $H = I, \lambda = 1/k$ and $\theta = 0$, (3.2.5) is the ridge regression estimator of Hoerl and Kennard (1970) derived in Section 1.3.

Example 3.2.1. Geometric Illustration of the Ridge Optimization Problem. A simple illustration of the procedure in Equations (3.2.1) to (3.2.5) above is finding the points on an ellipse or a circle in the plane closest to the origin. For instance, one could find the point of the circle $(X - 1)^2 + (Y - 1)^2 = \frac{1}{2}$ closest to the origin. To do this write

$$W = X^2 + Y^2 + \lambda[(X - 1)^2 + (Y - 1)^2 - \frac{1}{2}].$$

Differentiating,

$$W_X = 2X + 2\lambda(X - 1) = 0,$$

$$W_Y = 2Y + 2\lambda(Y - 1) = 0.$$

Now

$$\lambda = \frac{-X}{X - 1} = \frac{-Y}{Y - 1}$$

and the points at maximum or minimum distance lie on the circle and $Y = X$.

Thus,

$$2(X - 1)^2 = \frac{1}{2},$$

$$(X - 1) = \pm\frac{1}{2},$$

$$X = \frac{3}{2}, \frac{1}{2},$$

$$Y = \frac{3}{2}, \frac{1}{2},$$

so points at maximum or minimum distance from the origin are $(\frac{3}{2}, \frac{3}{2})$ and $(\frac{1}{2}, \frac{1}{2})$. Clearly, the desired point is $(\frac{1}{2}, \frac{1}{2})$.

For a linear regression model with two parameters and a data set where the LS estimators are $b_0 = 1$ and $b_1 = 1$, the ridge estimators would be $b_0 = \frac{1}{2}$ and $b_1 = \frac{1}{2}$ with $k = 1$.

Example 3.2.2. Special Case of Ridge Optimization Problem. Consider the linear model

$$
\begin{bmatrix} Y_{11} \\ Y_{12} \\ Y_{13} \\ Y_{14} \\ Y_{21} \\ Y_{22} \\ Y_{23} \\ Y_{24} \end{bmatrix} = \begin{bmatrix} 1_4 & 0 \\ 0 & 1_4 \end{bmatrix} \begin{bmatrix} \beta_1 \\ \beta_2 \end{bmatrix} + \epsilon.
$$

The LS estimators are $\hat{\beta}_1 = \bar{Y}_{1.}$, $\hat{\beta}_2 = \bar{Y}_{2.}$. The ridge estimators would have the form

$$
\hat{\beta}_1 = \frac{Y_{1.}}{4+k}, \qquad \hat{\beta}_2 = \frac{Y_{2.}}{4+k}.
$$

Consider the problem of minimizing $\beta_1^2 + \beta_1^2$ given

$$
(\beta_1 - \bar{Y}_{1.})^2 + (\beta_2 - \bar{Y}_{2.})^2 = C^2. \tag{3.2.6}
$$

Now let

$$
W = \beta_1^2 + \beta_2^2 + \lambda[4(\beta_1 - \bar{Y}_{1.})^2 + 4(\beta_2 - \bar{Y}_{2.})^2 - C^2].
$$

Differentiate, and obtain

$$
W_{\beta_1} = 2\beta_1 + 8\lambda(\beta_1 - \bar{Y}_{1.}) = 0
$$

and

$$W_{\beta_2} = 2\beta_2 + 8\lambda(\beta_2 - \bar{Y}_{2.}) = 0.$$

Thus,

$$\lambda = -\frac{\beta_1}{4(\beta_1 - \bar{Y}_{1.})} = -\frac{\beta_2}{4(\beta_2 - \bar{Y}_{2.})}.$$

Now $\beta_2 = \bar{Y}_{2.}/\bar{Y}_{1.}\beta_1$ may be substituted into (3.2.6) to obtain for $C > 0$ the "closest estimators",

$$\hat{\beta}_1 = \bar{Y}_{1.}\left(1 - \frac{C}{\sqrt{\bar{Y}_{1.}^2 + \bar{Y}_{2.}^2}}\right),$$

and

$$\hat{\beta}_2 = \bar{Y}_{2.}\left(1 - \frac{C}{\sqrt{\bar{Y}_{1.}^2 + \bar{Y}_{2.}^2}}\right).$$

The parameter

$$k = \frac{4C}{\sqrt{\bar{Y}_{1.}^2 + \bar{Y}_{2.}^2} - C}.$$

Exercise 3.2.1. Let

$$\begin{bmatrix} Y_{1j_1} \\ Y_{2j_2} \\ Y_{3j_3} \end{bmatrix} = \begin{bmatrix} 1_4 & 0 & 0 \\ 0 & 1_9 & 0 \\ 0 & 0 & 1_4 \end{bmatrix} \begin{bmatrix} \beta_0 \\ \beta_1 \\ \beta_2 \end{bmatrix} + \epsilon,$$

where $1 \le j_1 \le 4$, $1 \le j_2 \le 9$, $1 \le j_3 \le 4$.

A. Give the form of the least square estimator and ordinary ridge estimator.

B. Solve the optimization problem by the technique in equations (3.2.1) to (3.2.5) to obtain an expression for the ordinary ridge estimators and for k.

3.3 The Mixed Estimators

Consider the augmented model

$$\begin{bmatrix} Y \\ r \end{bmatrix} = \begin{bmatrix} X \\ R \end{bmatrix} \beta + \begin{bmatrix} \epsilon \\ \phi \end{bmatrix}, \tag{3.3.1}$$

where Y, X, β and ϵ are as defined in Section 1.2. The model

$$r = R\beta + \phi \tag{3.3.2}$$

may be thought of as taking some additional observations based on prior knowledge about the parameters or additional information. The matrix R is $t \times m$ and ϕ and r are t dimensional random variables. Let V_1 and V_2 be non-singular matrices.

The weighted LS estimator is obtained by minimizing

$$F_2(\beta) = (Y - X\beta)'V_1^{-1}(Y - X\beta) + (r - R\beta)'V_2^{-1}(r - R\beta). \tag{3.3.3}$$

This yields the normal equations

$$(X'V_1^{-1}X + R'V_2^{-1}R)\beta = X'V_1^{-1}Y + R'V_2^{-1}r. \tag{3.3.4}$$

The solution to (3.3.4) is the mixed estimator

$$p'\hat{\beta}m = p'(X'V_1^{-1}X + R'V_2^{-1}R)^+[X'V_1^{-1}Y + R'V_2^{-1}r]. \tag{3.3.5}$$

The parametric functions $p'\beta$, are independent of the choice of generalized inverse if, for some $n \times t$ dimensional vector c,

$$p' = c' \begin{bmatrix} X \\ R \end{bmatrix} = c_1'X + c_2'R; \tag{3.3.6}$$

$p'\beta$ is estimable for model (3.3.1).

If

$$p' = t'X \tag{3.3.7}$$

for some vector t, then $p'\beta$ is estimable with respect to the model

$$Y = X\beta + \epsilon. \tag{3.3.8}$$

This corresponds to $c_1 = t$ and $c_2 = 0$ in (3.3.6). Thus, estimability with respect to (3.3.8) implies estimability with respect to (3.3.1).
  Likewise if

$$p' = s'R \tag{3.3.9}$$

for some vector s, then $p'\beta$ is said to be estimable with respect to

$$r = R\beta + \phi. \tag{3.3.10}$$

This corresponds to $c_1 = 0$ and $c_2 = s$. Thus estimability with respect to (3.3.10) also implies estimability with respect to (3.3.1).
  However estimability with respect to (3.3.1) need not imply estimability with respect to (3.3.8) or (3.3.10). This may be illustrated with the following counterexample. Let

$$Y = \begin{bmatrix} 1 & 1 & 0 \\ 1 & 1 & 0 \\ 1 & 0 & 1 \\ 1 & 0 & 1 \\ 0 & 1 & 0 \\ 0 & 0 & 1 \end{bmatrix} \begin{bmatrix} \beta_1 \\ \beta_2 \\ \beta_3 \end{bmatrix} + \epsilon. \tag{3.3.11}$$

Here $X = \begin{bmatrix} 1 & 1 & 0 \\ 1 & 1 & 0 \\ 1 & 0 & 1 \\ 1 & 0 & 1 \end{bmatrix}$ and $R = \begin{bmatrix} 0 & 1 & 0 \\ 0 & 0 & 1 \end{bmatrix}$.

  The parametric function $\beta_1 + \beta_2 + \beta_3$ is estimable for the augmented model since the system

$$\begin{aligned} 1 &= t_1 + t_2 + t_3 + t_4, \\ 1 &= t_1 + t_2 + t_5, \\ 0 &= t_3 + t_4 + t_6 \end{aligned} \tag{3.3.12a}$$

has the solution $t_1 = t_2 = 1/2$, $t_3 = t_4 = t_5 = t_6 = 0$. The system

$$
\begin{aligned}
1 &= t_1 + t_2 + t_3 + t_4, \\
1 &= t_1 + t_2, \\
1 &= t_3 + t_4
\end{aligned}
\qquad (3.3.12b)
$$

has no solution and hence $\beta_1 + \beta_2 + \beta_3$ is not estimable for the model $Y = X\beta + \epsilon$.

The above discussion motivates the definition below.

Definition 3.3.1. Given an augmented model

$$
\begin{bmatrix} Y \\ r \end{bmatrix} = \begin{bmatrix} X \\ R \end{bmatrix} \beta + \begin{bmatrix} \epsilon \\ \phi \end{bmatrix},
$$

the parametric function $p'\beta$

1) is X estimable if it is estimable for the model

$$
Y = X\beta + \epsilon;
$$

2) is R estimable if it is estimable for the model

$$
r = R\beta + \phi;
$$

3) is (X, R) estimable if it is estimable for the augmented model.

An (X, R) estimable parametric function need not be X estimable or R estimable. However a parametric function that is either X estimable or R estimable is (X, R) estimable.

The estimator (3.3.4) is independent of the choice of generalized inverses if $p'\beta$ is X estimable, R estimable or (X, R) estimable.

Assume that for the model (3.3.1) the error terms ϵ and ϕ have mean zero and dispersions V_1 and V_2, i.e.,

$$
E(\epsilon) = 0, \qquad D(\epsilon) = V_1, \qquad (3.3.13a)
$$

$$E(\phi) = 0 \quad \text{and} \quad D(\phi) = V_2. \qquad (3.3.13b)$$

The matrices V_1 and V_2 are non-singular.

If $p'\beta$ is X estimable, R estimable, or (X, R) estimable then by the Gauss-Markov Theorem $p'\hat{\beta}_m$ is MVUE.

Remark. When V is of full rank and known, a model of the form

$$V^{-1/2}Y = V^{-1/2}X\beta + V^{-1/2}\epsilon \qquad (3.3.14)$$

with

$$E(V^{-1/2}\epsilon) = 0 \quad \text{and} \quad D(V^{-1/2}\epsilon) = I \qquad (3.3.15)$$

can be considered.

In the sequel the case of interest will be

$$E(\epsilon) = 0, \qquad D(\epsilon) = \sigma^2 I,$$

$$(3.3.16)$$

$$E(\phi) = 0, \qquad D(\phi) = \tau^2 I.$$

Example 3.3.1. Relationship between Ordinary Ridge Estimators and Mixed Estimators. The ordinary ridge regression estimator is a mixed estimator where $\sqrt{k}R$ is an orthogonal matrix and $V_1 = V_2 = \sigma^2 I$.

Example 3.3.2. A Mixed Estimator. Let

$$Y = \begin{bmatrix} 1_5 & 1_5 & 0 & 0 \\ 1_5 & 0 & 1_5 & 0 \\ 1_5 & 0 & 0 & 1_5 \end{bmatrix} \begin{bmatrix} \mu \\ \alpha_1 \\ \alpha_2 \\ \alpha_3 \end{bmatrix} + \epsilon.$$

Let

$$R = \begin{bmatrix} 0 & 1 & -1 & 0 \\ 0 & 1 & 1 & -2 \end{bmatrix}, \quad r = \begin{bmatrix} r_1 \\ r_2 \end{bmatrix}, \quad \text{and } V_1 = \omega^2 I = \tau^2 I = V_2.$$

Then the mixed estimator is

$$
\hat{\beta} = \left\{ \begin{bmatrix} 15 & 5 & 5 & 5 \\ 5 & 5 & 0 & 0 \\ 5 & 0 & 5 & 0 \\ 5 & 0 & 0 & 5 \end{bmatrix} + \begin{bmatrix} 0 & 0 \\ 1 & 1 \\ -1 & 1 \\ 0 & -2 \end{bmatrix} \begin{bmatrix} 0 & 1 & -1 & 0 \\ 0 & 1 & 1 & -2 \end{bmatrix} \right\}^{-}
$$

$$
\times \left\{ \begin{bmatrix} Y_{..} \\ Y_{1.} \\ Y_{2.} \\ Y_{3.} \end{bmatrix} + \begin{bmatrix} 0 \\ r_1 + r_2 \\ -r_1 + r_2 \\ -2r_2 \end{bmatrix} \right\}
$$

$$
= \begin{bmatrix} 15 & 5 & 5 & 5 \\ 5 & 7 & 0 & -2 \\ 5 & 0 & 7 & -2 \\ 5 & -2 & -2 & 9 \end{bmatrix}^{-} \begin{bmatrix} Y_{..} \\ Y_{1.} + r_1 + r_2 \\ Y_{2.} - r_1 + r_2 \\ Y_{3.} - 2r_2 \end{bmatrix}.
$$

One form of

$$
\hat{\beta} = \begin{bmatrix} \hat{\beta}_0 \\ \hat{\beta}_1 \\ \hat{\beta}_2 \\ \hat{\beta}_3 \end{bmatrix} = \frac{1}{385} \begin{bmatrix} 0 & 0 & 0 & 0 \\ 0 & 59 & 4 & 14 \\ 0 & 4 & 59 & 14 \\ 0 & 14 & 14 & 59 \end{bmatrix} \begin{bmatrix} Y_{..} \\ \frac{1}{2}(r_1 + r_2) + \bar{Y}_{1.} - \bar{Y}_{..} \\ \frac{1}{2}(-r_1 + r_2) + \bar{Y}_{2.} - \bar{Y}_{..} \\ \bar{Y}_{3.} - \bar{Y}_{..} \end{bmatrix}.
$$

Thus,

$$\hat{\beta}_0 = 0,$$

$$\hat{\beta}_1 = \frac{1}{385}[\frac{55}{2}r_1 + \frac{63}{2}r_2 + 59\bar{Y}_1. + 4\bar{Y}_2. + 14\bar{Y}_3. - 77\bar{Y}..],$$

$$\hat{\beta}_2 = \frac{1}{385}[-\frac{55}{2}r_1 + \frac{63}{2}r_2 + 59\bar{Y}_2. + 14\bar{Y}_3. - 77\bar{Y}..],$$

$$\hat{\beta}_3 = \frac{1}{385}[14r_2 + 14\bar{Y}_1. + 14\bar{Y}_2. - 77\bar{Y}..].$$

Example 3.3.3. Numerical Example of a Mixed Estimator.
One way to view stochastic prior information is as additional observations. Thus, a mixed estimator is really a weighted least square estimator where some of the observations came from sample information and some came from prior information.

An example is now given where two data sets are merged to form a weighted least square estimator. If, say, one of the data sets were collected at an earlier time it could be recorded as prior information.

The data below concerns student performance in a class taught by the author on Solutions to Engineering Problems. The course includes topics in Differential Equations and Multivariable Calculus. The dependent variable is the final examination score. The independent variables are the scores on four periodic hour examinations. The first data set Table 3.3.1 represents the prior information, the second data set the sample information.

Table 3.3.1 Prior Information

	Exam1	Exam2	Exam3	Exam4	Final
1	95	85	97	83	91
2	69	70	70	72	68
3	89	84	96	85	88
4	66	58	100	70	73
5	77	94	100	86	74
6	49	71	94	80	71

Table 3.3.2   Sample Information

	Test1	Test2	Test3	Test4	Final
1	81	82	100	78	68
2	89	86	100	73	95
3	90	96	97	88	95
4	87	75	93	83	61
5	74	71	88	73	71
6	64	65	89	67	39
7	89	88	93	64	63
8	56	63	83	34	42
9	93	86	92	85	95

Now for the first data set

$$b_1 = \begin{bmatrix} -38.7376 \\ 0.58677 \\ -1.01219 \\ -0.037448 \\ 1.94286 \end{bmatrix}.$$

For the second data set,

$$b_2 = \begin{bmatrix} -37.5502 \\ 0.30232 \\ 1.15323 \\ -0.29516 \\ 0.26935 \end{bmatrix}.$$

For the augmented model, using the reciprocal of Sum of Squares due to error divided by its degrees of freedom as weights, (Note σ^2 and τ^2 usually are not known.)

$$b_m = \begin{bmatrix} -8.5788 \\ .52007 \\ -.47379 \\ .11370 \\ .92344 \end{bmatrix}.$$

Here more weight is given to the "prior information." Just combining the two data sets, i.e. $\sigma^2 = \tau^2$,

$$b = \begin{bmatrix} -29.6748 \\ -.36683 \\ .236916 \\ .167127 \\ .53514 \end{bmatrix}.$$

Exercise 3.3.1. For the model in Exercise 3.1.1 let

$$R = \begin{bmatrix} 0 & 0 & -1 & 1 \\ 0 & -2 & 1 & 1 \end{bmatrix}, \qquad r = \begin{bmatrix} -1 \\ 2 \end{bmatrix}.$$

Write down the mixed estimator when $V_1 = 2I$, $V_2 = I$.

Exercise 3.3.2. For the data in Exercise 1.2.1 the vector r and matrix R result from taking two additional observations,

$$r = \begin{bmatrix} 2.9 \\ 3.8 \end{bmatrix} \qquad R = \begin{bmatrix} 1 & 1.6 & 2.5 \\ 1 & 2.4 & 5.0 \end{bmatrix}.$$

Write down the mixed estimator.

Exercise 3.3.3. For Example 3.3.2 which of the following parametric functions are (1) X estimable? (2) R estimable? (3) (X, R) estimable?

(1) α_2,
(2) $\alpha_1 - \alpha_2$,
(3) $\alpha_1 - \alpha_3$,
(4) $15\mu + 7\alpha_1 + 5\alpha_2 + 3\alpha_3$.

Exercise 3.3.4. For the models in Exercise 3.3.3 find another LS or mixed estimator. Show that the resulting parametric functions are the same.

Exercise 3.3.5. For the model in Exercise 3.1.1 and the exact constraint

$$-\alpha_2 + \alpha_3 = -1,$$

$$-2\alpha_1 + \alpha_2 + \alpha_3 = 2,$$

find the restricted LS estimator. [This problem may be solved for any exact constraint by minimizing

$$H = (Y - X\beta)'(Y - X\beta) + \lambda'(R'\beta - c)$$

where λ is a vector of Lagrange multipliers.]

Exercise 3.3.6. For the model in Example 3.3.2
   A. Find two LS estimators for the models with design matrices X, R, and $\begin{bmatrix} X \\ R \end{bmatrix}$.
   B. In each case when are these two LS estimators the same for each of the parametric functions in Exercise 3.3.3?

Exercise 3.3.7. Given the augmented linear model (3.3.1), let W be a scalar where $0 \le w \le 1$. Assume ϵ and ϕ satisfy (3.3.16). Minimize

$$L = (Y - X\beta)'(Y - X\beta)\frac{1 - w}{\sigma^2} + \frac{w}{\tau^2}(r - R\beta)'(r - R\beta)$$

to obtain the weighted mixed regression estimator (Toutenburg (1987)). What estimator do you get if $w = 0$? if $w = 1$?

Exercise 3.3.8. Formulate an optimization problem that will yield a "weighted ridge regression estimator".

Exercise 3.3.9. Same problem as Exercise 3.3.7 except minimize

$$L = (Y - X\beta)'(Y - X\beta)\frac{1 - w}{\sigma^2} + w(r - R\beta)'(r - R\beta).$$

[Here relative weights are given to sample and prior information.] What happens when $w = 0, \frac{1}{2}, 1$?

3.4 The Linear Minimax Estimator

Let β be an m dimensional vector of parameters. Let θ be an m dimensional vector. Let the parameter space

$$\Omega = \{\beta \ : \ (\beta - \theta)'G(\beta - \theta) \leq 1\}. \tag{3.4.1}$$

Assume G is positive definite. Consider a linear estimator of the form

$$p'\hat{\beta} = p'\theta + L'(Y - X\theta). \tag{3.4.2}$$

Using the methodology of Section 2.3:

1. The risk of (3.4.2) shall be found.
2. The maximum value of this risk on the ellipsoid in (3.4.1) will be obtained in terms of L.
3. The vector L so that the minimum value of the expression promised in 2 above is obtained will be found.

The risk or MSE is

$$p'Rp = p'E(\hat{\beta} - \beta)(\hat{\beta} - \beta)'p$$

$$= E[p'(\theta - \beta) + L'(Y - X\theta)][(\theta - \beta)'p + (Y' - \theta'X')L]$$

$$= (L'X - p')(\beta - \theta)(\beta - \theta)'(XL - p) + \sigma^2 L'L. \tag{3.4.3}$$

On the ellipsoid (3.4.1) or equivalently

$$(\beta - \theta)(\beta - \theta)' \leq G^{-1} = F, \tag{3.4.4}$$

the maximum of (3.4.3) is

$$p'Rp = (L'X - p')F(X'L - p) + \sigma^2 L'L. \qquad (3.4.5)$$

Differentiating with respect to L,

$$2XFX'L - 2p'FX' + L\sigma^2 = 0. \qquad (3.4.6)$$

Thus, since XFX' is NND, the expression in (3.4.6) is minimized when

$$L' = p'FX'(XFX' + \sigma^2 I)^{-1}. \qquad (3.4.7)$$

Thus, the minimax estimator is

$$p'\hat{\beta} = p'\theta + p'FX(XFX' + \sigma^2 I)^{-1}(Y - X\theta). \qquad (3.4.8)$$

When G is of non-full rank the above results are valid if the ellipsoid is replaced by the region

$$(\beta - \theta)(\beta - \theta)' \le F. \qquad (3.4.9)$$

The matrix F is non-negative definite and need not be of full rank. If $(\beta - \theta)$ is in the range of F then from, Theorem 2.5.3, equation (3.4.9) is equivalent to an ellipsoid of the form (3.4.1).

When $U'GU$ is positive definite then

$$(\beta - \theta)'(UU'GUU')^+(\beta - \theta) \le 1 \qquad (3.4.10)$$

iff

$$U'(\beta - \theta)(\beta - \theta)'U \le U'GU. \qquad (3.4.11)$$

When $p'\beta$ is estimable $p'UU' = p'$. Also $X = XUU'$. Equation (3.4.3) may then be written

$$p'Rp = (L'X - p)UU'(\beta - \theta)(\beta - \theta)'UU'(XL - p) + \sigma^2 L'L. \qquad (3.4.12)$$

The maximum of $p'Rp$ on the ellipsoid in (3.4.10) is (3.4.5). This can then be minimized to obtain the minimax estimator (3.4.8).

Example 3.4.1. A Minimax Estimator. Consider the linear model

$$Y_{1j} = \mu_1 + \epsilon_1, \quad 1 \le j \le n_1,$$

$$Y_{2j} = \mu_2 + \epsilon_2, \quad 1 \le j \le n_2,$$

with

$$E(\epsilon_i) = 0, \quad D(\epsilon_1) = \sigma^2.$$

Find the estimator of $\mu_1 - \mu_2$ of the form $c_1 \bar{Y}_{1.} + c_2 \bar{Y}_{2.}$ whose maximum on $\mu_1^2 + \mu_2^2 \le 1$ minimizes

$$g(c_1, c_2) = E[c_1 \bar{Y}_{1.} + c_2 \bar{Y}_{2.} - (\mu_1 - \mu_2)]^2.$$

Observe that

$$g(c_1, c_2) = \frac{c_1^2 \sigma^2}{n_1} + \frac{c_2^2 \sigma^2}{n_2} + [(c_1 - 1)\mu_1 + (c_2 + 1)\mu_2]^2.$$

Now (μ_1, μ_2) lies on the disk $\mu_1^2 + \mu_2^2 \le 1$ iff

$$\begin{bmatrix} \mu_1 \\ \mu_2 \end{bmatrix} [\mu_1, \mu_2] \le I$$

by Theorem 2.5.2. Thus, when (μ_1, μ_2) lies on the unit disk

$$g(c_1, c_2) = \frac{c_1^2 \sigma^2}{n_1} + \frac{c_2^2 \sigma^2}{n_2}$$

$$+ [c_1 - 1, \quad c_2 + 1] \begin{bmatrix} \mu_1 \\ \mu_2 \end{bmatrix} [\mu_1 \quad \mu_2] \begin{bmatrix} c_1 - 1 \\ c_2 + 1 \end{bmatrix}$$

$$\le \frac{c_1^2 \sigma^2}{n_1} + \frac{c_2^2 \sigma^2}{n_2} + (c_1 - 1)^2 + (c_2 + 1)^2$$

$$= \max g(c_1, c_2) = h(c_1, c_2).$$

To minimize $h(c_1, c_2)$

$$\frac{\partial h}{\partial c_1} = \frac{2c_1\sigma^2}{n_1} + 2(c_1 - 1) = 0$$

and

$$\frac{\partial h}{\partial c_2} = \frac{2c_2\sigma^2}{n_2} + 2(c_2 + 1) = 0.$$

Now

$$c_1 = \frac{1}{1 + \frac{\sigma^2}{n_1}} \quad \text{and} \quad c_2 = -\frac{1}{1 + \frac{\sigma^2}{n_2}}.$$

Furthermore

$$\frac{\partial^2 h}{\partial c_1^2} = \frac{2\sigma^2}{n_1} + 2, \qquad \frac{\partial^2 h}{\partial c_1 \partial c_2} = 0$$

and

$$\frac{\partial^2 h}{\partial c_2^2} = \frac{2\sigma^2}{n_2} + 2,$$

so the minimum estimator is of the form

$$\hat{\mu}_1 - \hat{\mu}_2 = \frac{1}{1 + \frac{\sigma^2}{n_1}} \bar{Y}_{1.} - \frac{1}{1 + \frac{\sigma^2}{n_2}} \bar{Y}_{2.} .$$

Exercise 3.4.1. Consider the model in Example 3.4.1 above. Find the minimax estimator of $2\mu_1 + \mu_2$ given the ellipsoid $\mu_1^2/9 + \mu_2^2/4 \leq 1$.

3.5 The Bayes Estimator

Assume that β is a random variable with known prior mean and known dispersion given by

$$E(\beta) = \theta \quad \text{and} \quad D(\beta) = F. \tag{3.5.1}$$

Assume that

$$E(\epsilon|\beta) = 0 \quad \text{and} \quad D(\epsilon|\beta) = \sigma^2 I. \tag{3.5.2}$$

As in C.R. Rao (1973) minimize the variance

$$v = \text{Var}(p'\beta - a - L'Y)$$

$$= \text{Var}(p'\beta) + \text{Var}(L'Y) - 2\text{cov}(p'\beta,\ L'Y) \tag{3.5.3a}$$

$$= p'Fp + L'(XFX' + \sigma^2 I)L - 2p'FX'L$$

subject to

$$E(p'\beta - a - L'Y) = 0. \tag{3.5.3b}$$

Equation (3.5.3b) implies that

$$a = (p' - L'X)\theta. \tag{3.5.4}$$

Differentiate (3.5.3) to obtain

$$(XFX' + \sigma^2 I)L = XFp. \tag{3.5.5}$$

Thus,

$$L' = FX'(XFX' + \sigma^2 I)^{-1}. \tag{3.5.6}$$

Notice that $XFX' + \sigma^2 I$ is positive definite. The optimum estimator for both the full and the non-full rank case is, from Equations (3.5.4), (3.5.6) and the discussion above,

$$p'\hat{\beta}_b = p'\theta + p'FX'(XFX' + \sigma^2 I)^{-1}(Y - X\theta). \qquad (3.5.7)$$

Notice that (3.5.7) is the MVUE relative to the assumptions in (3.5.1) and (3.5.2). A Gauss-Markov theorem will be given in the next chapter.

When θ is unknown and $p' = L'X$, i.e., $p'\beta$ is estimable, expression (3.5.3) reduces to

$$v = L'L\sigma^2. \qquad (3.5.8)$$

Minimize v in (3.5.8) together with $p' = L'X$. The proof of the Gauss-Markov Theorem yields the LS estimator.

A modified form of the optimization problem in (3.5.3) is to minimize

$$v = \text{Var}(p'\beta - a - c'b) \qquad (3.5.9)$$

with a constant and c a vector, subject to

$$E(p'\beta - a - c'b) = 0. \qquad (3.5.10)$$

Now

$$v = \text{Var}(p'\beta) + \text{Var}(c'b) - 2\,\text{cov}(p'\beta,\, c'b)$$

$$\qquad (3.5.11)$$

$$= p'Fp + c'[UU'FUU' + \sigma^2(X'X)^+]c - 2p'FUU'c.$$

Equation (3.5.10) implies that

$$p'\theta - c'UU'\theta = a. \qquad (3.5.12)$$

Differentiating (3.5.11),

$$[UU'FUU' + \sigma^2(X'X)^+]c = p'FUU'. \qquad (3.5.13)$$

Now $UU'FUU' + \sigma^2(X'X)^+$ is NND. From (3.5.12) and (3.5.13),

$$p'\hat{\beta} = p'\theta + p'FUU'[UU'FUU' + \sigma^2(X'X)^+]^+(b - \theta) \quad (3.5.14)$$

is obtained. Conditions for (3.5.7) and (3.5.14) to be equivalent will be given in Chapter IV.

When θ is unknown (3.5.12) becomes the condition $p' = c'UU'$ and $p'\beta$ is estimable. The problem then becomes one of minimizing

$$v = c'(X'X)^+c\sigma^2 \quad (3.5.15)$$

subject to

$$p' - c'UU' = 0. \quad (3.5.16)$$

Let λ be a vector of Lagrange multipliers. Let

$$T = c'(X'X)^+c - (p' - c'UU')\lambda. \quad (3.5.17)$$

Differentiate T with respect to c' and set the result equal to zero and obtain

$$(X'X)^+c + UU'\lambda = 0. \quad (3.5.18)$$

Multiply (3.5.18) by $X'X$. Then

$$UU'c + X'X\lambda = 0 \quad (3.5.19)$$

and

$$\lambda = -(X'X)^+UU'c = -(X'X)^+p \quad (3.5.20)$$

since $(X'X)^+UU' = U\Lambda^{-1}U'UU' = (X'X)^+$.

Substitute (3.5.20) in (3.5.18) to get $c = p$. Thus b is the solution and an alternative proof of the Gauss-Markov Theorem has been obtained as a byproduct of the above investigation.

The optimization problem above may be solved for any prior distribution with mean and variance given in (3.5.1). However (3.5.7), the linear BE, is the BE in the more general sense described in Section 2.2 if and only if the prior distribution is multivariate normal (see Kagan, Linnik, and Rao (1973)).

Example 3.5.1. A BE. Consider the linear model in Example 3.4.1. The problem is to find the BE of $\mu_1 - \mu_2$ in the form $c_1 \hat{Y}_1. + c_2 \hat{Y}_2.$, given $E(\mu_i) = 0$ and $\text{Var}(\mu_i) = 1$, where $i = 1, 2$. The μ_1 and μ_2 are independent.

Observe that

$$E(c_1 \bar{Y}_1. + c_2 \bar{Y}_2. - a_0 - (\mu_1 - \mu_2)) = -a_0 = 0.$$

Thus, it suffices to minimize

$$g(c_1, c_2) = \text{Var}(c_1 \bar{Y}_1. + c_2 \bar{Y}_2. - (\mu_1 - \mu_2))$$

$$= c_1^2 \ \text{Var}(\bar{Y}_1.) + c_2^2 \ \text{Var}(\bar{Y}_2.) + \text{Var}(\mu_1) + \text{Var}(\mu_2)$$

$$- 2c_1 \text{cov}(\bar{Y}_1., \mu_1) + 2c_2 \text{cov}(\bar{Y}_2., \mu_2).$$

Set

$$\frac{\partial g}{\partial c_1} = 2c_1 \text{Var}(\bar{Y}_1.) - 2 \ \text{cov}(\bar{Y}_1., U_1) = 0$$

and

$$\frac{\partial g}{\partial c_2} = 2c_2 \text{Var}(\bar{Y}_2.) + 2 \ \text{cov}(\bar{Y}_2., U_2) = 0.$$

Then

$$c_1 = \frac{\text{cov}(\bar{Y}_1., \mu_1)}{\text{Var}(\bar{Y}_1.)} = \frac{1}{\frac{\sigma^2}{n_1} + 1}$$

and

$$c_2 = -\frac{\text{cov}(\bar{Y}_{2.}, \mu_2)}{\text{Var}(\bar{Y}_{2.})} = -\frac{1}{\frac{\sigma^2}{n_2} + 1}.$$

Thus,

$$(\mu_1 \hat{-} \mu_2)^b = \frac{1}{\frac{\sigma^2}{n_1} + 1}\bar{Y}_{1.} - \frac{1}{\frac{\sigma^2}{n_2} + 1}\bar{Y}_{2..}$$

This is the same result as that obtained in Example 3.4.1.

Exercise 3.5.1. For the above model find the BE of $2\mu_1 + \mu_2$ when $E(\mu_1) = 0$ and $D\begin{bmatrix} \mu_1 \\ \mu_2 \end{bmatrix} = \begin{bmatrix} \frac{1}{9} & 0 \\ 0 & \frac{1}{4} \end{bmatrix}$ Compare your answer to that of Exercise 3.4.1.

Exercise 3.5.2. Given the linear model

$$Y_i = \mu + \epsilon_i, \qquad 1 \le i \le n,$$

where $E(\epsilon_i) = 0$, $V(\epsilon_i) = \sigma^2 I$, together with the prior information about μ,

$$E(\mu) = 0, \qquad D(\mu) = \tau^2.$$

Minimize

$$v = E(c\bar{Y} - d - \mu)$$

given that

$$E(c\bar{Y} - d - \mu) = 0.$$

Thus, obtain the linear BE.

Exercise 3.5.3. Given the model (3.1) together with the prior assumptions (3.5.1) and assumptions (3.5.2). Minimize

$$v = \mathrm{Var}(p'\beta - a - L'X'Y)$$

where L is an m dimensional column vector subject to

$$E(p'\beta - a - L'X'Y) = 0.$$

Obtain the resulting optimum estimator. Under what conditions is it equivalent to (3.5.7)? (3.5.14)?

3.6 Summary and Remarks

The following was accomplished in this chapter:

1. The concept of estimability was presented as a criterion for the uniqueness of LS estimators.
2. It was shown that the LS estimator was the minimum variance unbiased estimator for the estimable parametric functions.
3. The generalized ridge estimator, the mixed estimator, the minimax estimator and the linear Bayes estimator were each obtained by solving a different optimization problem.

Several things are apparent:

1. The algebraic form of the mixed estimator and the generalized ridge regression estimator are similar.
2. The form of the Bayes estimator and the minimax estimator are exactly the same.

The precise relationship between these estimators is the subject of the next chapter.

Chapter IV

How the Different Estimators Are Related

4.0 Introduction

The Bayes, generalized ridge mixed and minimax estimators have similar mathematical forms. The conditions for these estimators to be equivalent will now be explored.

C.R. Rao (1975) obtained different forms of the Bayes estimator when X and F have full rank. These forms are useful for the following reasons.

1. Ridge type estimators may be derived as special cases of Bayes estimators.
2. The relationship between the mixed estimator and the Bayes estimator can be studied.
3. The relative weight of sample and prior information can be considered.
4. James-Stein type estimators can be derived as special cases of empirical Bayes estimators.

Section 4.1 is concerned with these alternative forms.

The results contained in Section 4.1 are generalized in Section 4.2 to situations when X and F need not be of full rank. The case where $U'FU$ is positive definite is particularly interesting because the ridge type estimators in the literature satisfy this condition. Examples of situations where $U'FU$ is not positive definite have been mentioned by Theil (1971), Swamy, Mehta, Thurman and Iyengar (1985) and Thurman, Swamy and Mehta (1986). These are taken up in Section 4.6.

The conditions when the generalized ridge regression estimator is equivalent to the Bayes estimator are derived in Section 4.3.

In Section 4.4 conditions are obtained for sample and prior information to be exchangeable. Under conditions described there the roles of the design matrix and the prior dispersion matrix may be interchanged.

How the ridge regression estimators in the literature fit into the general picture is taken up in Section 4.5.

Finally a generalization of the Gauss-Markov Theorem is given at the end of Section 4.6. This theorem was studied by Duncan and Horn (1972) for the Kalman filter and by Harville (1976) for the variance components model. The theorem is stated in the context of the mathematical developments of this and the previous chapter. In Chapters IX and X it will be taken up again in the context of the Kalman filter and the variance components model.

4.1 Alternative Forms of the Bayes Estimator Full Rank Case

Five alternative forms of the Bayes estimator will be introduced in Theorem 4.1.1 below. These forms will be helpful in understanding the relationships between the Bayes, mixed and generalized ridge regression estimators introduced in Chapter III. In this section the full rank case is considered. Section 4.2 deals with the non-full rank case. These forms were originally obtained for the full rank case by C.R. Rao (1975).

Theorem 4.1.1. Let X be of full rank and F be a positive definite matrix. Then five alternative forms of the BE are

$$p'\hat{\beta} = p'\theta + p'[X'X + \sigma^2 F^{-1}]^{-1} X'(Y - X\theta) \qquad (4.1.1)$$

$$= p'[X'X + \sigma^2 F^{-1}]^{-1}[X'Y + F^{-1}\theta\sigma^2] \tag{4.1.2}$$

$$= p'\theta + p'F[\sigma^2(X'X)^{-1} + F]^{-1}(b - \theta) \tag{4.1.3}$$

$$= p'(X'X)^{-1}[F + \sigma^2(X'X)^{-1}]^{-1}\theta\sigma^2$$

$$+ p'F[F + \sigma^2(X'X)^{-1}]^{-1}b \tag{4.1.4}$$

$$= p'b - p'(X'X)^{-1}[F + \sigma^2(X'X)^{-1}]^{-1}(b - \theta)\sigma^2. \tag{4.1.5}$$

Before establishing the above results their usefulness will be considered.

When $\theta = 0$ the first form is the generalized ridge estimator of C.R. Rao (1975). It takes the form

$$p'\hat{\beta} = p'[X'X + G]^{-1}X'Y, \tag{4.1.6}$$

where $G = \sigma^2 F^{-1}$. For various choices of G, e.g. $G = kI$, different ridge estimators result (see Section 4.5).

The second form, Equation (4.1.2), will be used to

1. develop an extended Gauss-Markov theorem;
2. establish a relationship between mixed estimators and Bayes estimators.

Equation (4.1.3), the third form gives

1. an expression for the BE in terms of the OLS;
2. the solution to the second optimization problem of Section 3.5.

The relative importance of the sample and the prior information is expressed in the fourth form (4.1.4). Equation (4.1.4) is a weighted average of θ and b.

Equation (4.1.5) is used by C.R. Rao (1975) and Gruber (1979) to formulate empirical Bayes estimators. Empirical Bayes estimators are formulated when σ^2, F and θ are unknown or not completely known by substituting sample estimators for functions of σ^2, F and θ.

The proof of Theorem 4.1 will require the matrix identity

$$(A + BCD)^{-1} = A^{-1} - A^{-1}B(C^{-1} + DA^{-1}B)^{-1}DA^{-1}. \tag{4.1.7}$$

(See, for example, Toutenburg (1982).)

Proof of Theorem 4.1.1.

1. Verification of Equation 4.1.1.
   In Equation 4.1.7 above let $A = \sigma^2 I$, $B = X$, $C = F$ and $D = X$.
Thus,

$$(\sigma^2 I + X F X')^{-1} = \frac{1}{\sigma^2} I - \frac{1}{\sigma^2} X (F^{-1} + \frac{1}{\sigma^2} X' X)^{-1} \frac{X'}{\sigma^2}. \quad (4.1.8)$$

Consequently

$$p'\hat{\beta} = p'\theta + p' F X' (X F X' + \sigma^2 I)^{-1} (Y - X\theta)$$

$$= p'\theta + p' F X' [\frac{I}{\sigma^2} - \frac{1}{\sigma^2} X (F^{-1} + \frac{1}{\sigma^2} X' X)^{-1} \frac{X'}{\sigma^2}] (Y - X\theta)$$

$$= p'\theta + p' F X' (Y - X\theta) p'$$

$$- \frac{F X' X}{\sigma^2} (F^{-1} + \frac{1}{\sigma^2} X' X)^{-1} \frac{X'}{\sigma^2} (Y - X\theta).$$

$$(4.1.9)$$

Now

$$I - \sigma^2 F^{-1} (\sigma^2 F^{-1} + X' X)^{-1} = X' X (\sigma^2 F^{-1} + X' X)^{-1}. \quad (4.1.10)$$

Substitute the lefthand side of (4.1.10) in (4.1.9) for $X' X (\sigma^2 F^{-1} + X' X)^{-1}$ to obtain (4.1.1).

2. Verification of Equation 4.1.2.
   Equation 4.1.2 may be obtained from (4.1.1) by noticing that

$$I - [X' X + \sigma^2 F^{-1}]^{-1} X' X = [X' X + \sigma^2 F^{-1}]^{-1} \sigma^2 F^{-1}. \quad (4.1.11)$$

3. Verification of Equation 4.1.3.

Notice that

$$(F^{-1}\sigma^2 + X'X)^{-1}X'X$$

$$= [X'X[(X'X)^{-1}\sigma^2 + F]F^{-1}]^{-1}X'X \qquad (4.1.12)$$

$$= F[(X'X)^{-1}\sigma^2 + F]^{-1}.$$

From (4.1.1), since $(X'X)(X'X)^{-1} = I$,

$$p'\hat{\beta} = p'\theta + p'[X'X + \sigma^2 F^{-1}]^{-1}(X'X)(X'X)^{-1}X'(Y - X\theta). \quad (4.1.13)$$

Substitute (4.1.12) into (4.1.13) to obtain (4.1.3).

4. Verification of Equation 4.1.4.
   Equation (4.1.4) follows from (4.1.3) since

$$I - F[\sigma^2(X'X)^{-1} + F]^{-1} = \sigma^2(X'X)^{-1}[\sigma^2(X'X)^{-1} + F]^{-1}. \quad (4.1.14)$$

5. Verification of Equation 4.1.5.
   Equation (4.1.5) follows from (4.1.4) by (4.1.14). ∎

Example 4.1.1. Different Forms of the BE. Let

$$\begin{bmatrix} Y_1 \\ Y_2 \\ Y_3 \end{bmatrix} = \begin{bmatrix} -2 & 1 & 0 \\ 1 & 1 & -1 \\ 1 & 1 & 1 \end{bmatrix} \begin{bmatrix} \beta_1 \\ \beta_2 \\ \beta_3 \end{bmatrix} + \begin{bmatrix} \epsilon_1 \\ \epsilon_2 \\ \epsilon_3 \end{bmatrix}.$$

The reader should verify that the least square estimators are

$$b_1 = \frac{1}{6}(-2Y_1 + Y_2 + Y_3),$$

$$b_2 = \frac{1}{3}(Y_1 + Y_2 + Y_3),$$

$$b_3 = \frac{1}{2}(-Y_2 + Y_3).$$

Assume

$$E(\epsilon) = 0 \quad \text{and} \quad D(\epsilon) = I.$$

Also

$$E(\beta) = \begin{bmatrix} \theta_1 \\ \theta_2 \\ \theta_3 \end{bmatrix} \quad \text{and} \quad D(\beta) = \begin{bmatrix} \frac{1}{6} & 0 & 0 \\ 0 & \frac{1}{3} & 0 \\ 0 & 0 & \frac{1}{2} \end{bmatrix}.$$

Now

$$FX' = \begin{bmatrix} \frac{1}{6} & 0 & 0 \\ 0 & \frac{1}{3} & 0 \\ 0 & 0 & \frac{1}{2} \end{bmatrix} \begin{bmatrix} -2 & 1 & 1 \\ 1 & 1 & 1 \\ 0 & -1 & 1 \end{bmatrix} = \begin{bmatrix} -\frac{1}{3} & \frac{1}{6} & \frac{1}{6} \\ \frac{1}{3} & \frac{1}{3} & \frac{1}{3} \\ 0 & -\frac{1}{2} & \frac{1}{2} \end{bmatrix},$$

$$(XFX' + I)^{-1} = \begin{bmatrix} \frac{1}{2} & 0 & 0 \\ 0 & \frac{1}{2} & 0 \\ 0 & 0 & \frac{1}{2} \end{bmatrix}$$

and

$$FX'(XFX' + I)^{-1} = \begin{bmatrix} -\frac{1}{6} & \frac{1}{12} & \frac{1}{12} \\ \frac{1}{6} & \frac{1}{6} & \frac{1}{6} \\ 0 & -\frac{1}{4} & \frac{1}{4} \end{bmatrix}.$$

Forms of the BE are

$$
\begin{bmatrix} \hat{\beta}_1 \\ \hat{\beta}_2 \\ \hat{\beta}_3 \end{bmatrix} = \begin{bmatrix} \theta_1 \\ \theta_2 \\ \theta_3 \end{bmatrix} + \begin{bmatrix} -\frac{1}{6} & \frac{1}{12} & \frac{1}{12} \\ \frac{1}{6} & \frac{1}{6} & \frac{1}{6} \\ 0 & -\frac{1}{4} & \frac{1}{4} \end{bmatrix} \begin{bmatrix} Y_1 + 2\theta_1 - \theta_2 \\ Y_2 - \theta_1 - \theta_2 + \theta_3 \\ Y_3 - \theta_1 - \theta_2 - \theta_3 \end{bmatrix}
$$

$$
= \begin{bmatrix} \theta_1 \\ \theta_2 \\ \theta_3 \end{bmatrix} + \begin{bmatrix} \frac{1}{12} & 0 & 0 \\ 0 & \frac{1}{6} & 0 \\ 0 & 0 & \frac{1}{4} \end{bmatrix} \begin{bmatrix} -2Y_1 + Y_2 - 6\theta_1 \\ Y_1 + Y_2 - Y_3 - 3\theta_2 \\ Y_1 + Y_2 + Y_3 - 2\theta_3 \end{bmatrix}
$$

$$
= \begin{bmatrix} \frac{1}{12} & 0 & 0 \\ 0 & \frac{1}{6} & 0 \\ 0 & 0 & \frac{1}{4} \end{bmatrix} \begin{bmatrix} -2Y_1 + Y_2 \\ Y_1 + Y_2 - Y_3 \\ Y_1 + Y_2 - Y_3 \end{bmatrix} + \begin{bmatrix} 6 & 0 & 0 \\ 0 & 3 & 0 \\ 0 & 0 & 2 \end{bmatrix} \begin{bmatrix} \theta_1 \\ \theta_2 \\ \theta_3 \end{bmatrix}
$$

$$
= \begin{bmatrix} \theta_1 \\ \theta_2 \\ \theta_3 \end{bmatrix} + \begin{bmatrix} \frac{1}{2} & 0 & 0 \\ 0 & \frac{1}{2} & 0 \\ 0 & 0 & \frac{1}{2} \end{bmatrix} \begin{bmatrix} b_1 - \theta_1 \\ b_2 - \theta_2 \\ b_3 - \theta_3 \end{bmatrix}
$$

$$
= \begin{bmatrix} \frac{1}{2} & 0 & 0 \\ 0 & \frac{1}{2} & 0 \\ 0 & 0 & \frac{1}{2} \end{bmatrix} \begin{bmatrix} \theta_1 \\ \theta_2 \\ \theta_3 \end{bmatrix} + \begin{bmatrix} \frac{1}{2} & 0 & 0 \\ 0 & \frac{1}{2} & 0 \\ 0 & 0 & \frac{1}{2} \end{bmatrix} \begin{bmatrix} b_1 \\ b_2 \\ b_3 \end{bmatrix}
$$

$$
= \begin{bmatrix} b_1 \\ b_2 \\ b_3 \end{bmatrix} - \begin{bmatrix} \frac{1}{2} & 0 & 0 \\ 0 & \frac{1}{2} & 0 \\ 0 & 0 & \frac{1}{2} \end{bmatrix} \begin{bmatrix} b_1 - \theta_1 \\ b_2 - \theta_1 \\ b_3 - \theta_1 \end{bmatrix}.
$$

If $\theta = \begin{bmatrix} -1 \\ 2 \\ 2 \end{bmatrix}$, then

$$2\hat{\beta}_1 - \hat{\beta}_2 - \hat{\beta}_3 = -\frac{3}{4}Y_1 - \frac{1}{4}Y_2 - \frac{1}{12}Y_3 - 3$$

$$= 2(\frac{1}{2}b_1 - \frac{1}{2}) - \frac{1}{2}(b_2 + 1) - \frac{1}{2}(b_3 + 1)$$

$$= b_1 - \frac{1}{2}b_2 - \frac{1}{2}b_3 - 2.$$

Exercise 4.1.1. Verify the equalities above.

Exercise 4.1.2. Let

$$Y_i = \mu + \epsilon_i, \qquad 1 \le i \le n,$$

$$E(\epsilon_i|\mu) = 0, \qquad D(\epsilon_i|\mu) = \sigma^2,$$

$$E(\mu) = \theta, \qquad D(\mu) = \tau^2.$$

Assume ϵ_i are independent and ϵ_i and μ are independent. What are the different alternative forms of the BE?

4.2 Alternative Forms of the Bayes Estimator Non-Full Rank Case

In this section the five alternative forms of the BE will be presented for the non-full rank case—see Equations (4.2.4), (4.2.7), (4.2.8) and (4.2.9) below. The forms are generally the same as those for the full rank case with $UU'FUU'$ replacing F and Moore-Penrose inverses replacing the ordinary inverses.

For the non-full rank case the results of this section will help

1. develop an extended GM theorem;
2. show how ridge type estimators are special cases of Bayes estimators;
3. clarify the relationship between the mixed estimators and the BE.

For some of the equivalences to hold either a condition on the prior mean or dispersion will have to be imposed or the parametric functions will have to be estimable. The derivation of the BE when θ is known is valid for both estimable and non-estimable parametric functions, because it was done for all parametric functions. On the other hand, the derivation of the optimum estimator for an unknown prior mean was done assuming that the parametric functions were estimable. The result was the least square estimator.

Once the BE have been derived for all parametric functions (estimable or not) the condition of estimability may be imposed, if needed, to obtain equivalent estimators.

4.2.1 First Alternative Form

Notice that

$$X = S\Lambda^{1/2}U' = S'\Lambda^{1/2}U'UU' = XUU'. \qquad (4.2.1)$$

Also

$$
\begin{aligned}
(\sigma^2 I &+ XFX')^{-1} \\
&= (\sigma^2 I + XUU'FUU'X')^{-1} \\
&= \frac{1}{\sigma^2}I - \frac{1}{\sigma^2}XU[\frac{U'X'XU}{\sigma^2} + (U'FU)^{-1}]^{-1}\frac{U'X'}{\sigma^2},
\end{aligned}
\qquad (4.2.2)
$$

provided that $(U'FU)$ is positive definite.
Substituting (4.2.2) into (3.5.7),

$$p'\hat{\beta} = p'\theta + p'FU(U'FU)^{-1}[\frac{U'X'XU}{\sigma^2} + (U'FU)^{-1}]^{-1}\frac{U'X'}{\sigma^2}(Y - X\theta) \qquad (4.2.3)$$

is obtained.

Since $UU'+VV' = I$, when $p'\hat{\beta}$ is estimable or $V'FU = 0$ Equation (4.2.3) may be written

$$p'\hat{\beta} = p'\theta + p'[X'X + \sigma^2(UU'FUU')^+]^+X'(Y - X\theta). \qquad (4.2.4)$$

4.2.2 Second Alternative Form

Observe that
$$I - [X'X + \sigma^2(UU'FUU')^+]^+X'X$$

$$= UU' + VV' - U[\Lambda + \sigma^2(U'FU)^{-1}]^{-1}\Lambda U'$$

$$= U[I - [\Lambda + \sigma^2(U'FU)^{-1}]^{-1}\Lambda]U' \qquad (4.2.5)$$

$$= U[\Lambda + \sigma^2(U'FU)^{-1}]^{-1}(U'FU)^{-1}U'\sigma^2$$

$$= [X'X + \sigma^2(UU'FUU')^+]^+(UU'FUU')^+\sigma^2.$$

Then (4.2.4) may be written
$$p'\hat{\beta} = p'[X'X+\sigma^2(UU'FUU')^+]^+[X'Y+(UU'FUU')^+\sigma^2\theta]. \quad (4.2.6)$$
As will be seen in Section 4.3 this is a form of the generalized ridge regression estimator.

Thus, (4.2.4) is equivalent to (3.5.7) if

1. the matrix $U'FU$ is positive definite;
2. either $p'\beta$ is estimable or $V'FU = 0$.

Given condition 2, it can be shown that condition 1 is necessary and sufficient for equivalence of (3.5.7) and (4.2.4).

4.2.3 Expressions in Terms of Least Square Estimator

Let $C = \sigma^2\Lambda^{-1} + U'FU$. From (4.2.3)
$$p'\hat{\beta} = p'\theta + p'FU(U'FU)^{-1}[\Lambda C(U'FU)]^{-1}U'X'(Y - X\theta)$$
$$= p'\theta + p'FUC^{-1}\Lambda^{-1}U'X'(Y - X\theta) \qquad (4.2.7)$$
$$= p'\theta + p'F[(X'X)^+\sigma^2 + (UU'FUU')]^+ (b - \theta).$$

Notice that estimability was not required. Only $U'FU$ being positive definite was needed. Notice that (4.2.7) is the solution to the optimization problems in Section 3.5.

Also the next alternative form is restated from (3.5.14):

$$p'\hat{\beta} = p'(X'X)^+[UU'FUU' + \sigma^2(X'X)^+]^+\theta\sigma^2$$

$$\text{(4.2.8)}$$

$$+ p'UU'FUU'[UU'FUU' + \sigma^2(X'X)^+]^+b.$$

Notice that

$$UU'FUU'[UU'FUU' + \sigma^2(X'X)^+]^+$$

$$= UU'FUU'U(U'FU + \sigma^2\Lambda^{-1})^{-1}U'$$

$$= UU'FU(U'FU + \sigma^2\Lambda^{-1})^{-1}U' \qquad \text{(4.2.9)}$$

$$= UU' - U\Lambda^{-1}(U'FU + \sigma^2\Lambda^{-1})^{-1}U'\sigma^2$$

$$= UU' - (X'X)^+[UU'FUU' + \sigma^2(X'X)^+]^+\sigma^2.$$

Since $UU'b = b$, (4.2.8) may now be written in a new alternative form

$$p'\hat{\beta} = p'b - p'(X'X)^+[UU'FUU' + \sigma^2(X'X)^+]^+(b - \theta)\sigma^2. \quad \text{(4.2.10)}$$

4.2.4 Some Remarks Concerning the Alternative Forms of the BE

The remarks of Section 3.1 concerning the importance of these forms are reiterated.

The first two forms (4.2.4) and (4.2.6) are of the form $a + L'Y$ and are MVUE of θ. The last three forms are the MVUE of the forms $a + c'b$. All of the forms are equivalent provided that $p'\beta$ is estimable and $U'FU$ is PD.

The forms (4.2.4) and (4.2.6) are interesting because they are special cases of ridge regression estimators. Equation (4.2.8) is a weighted average of the relative contribution of sample and prior information. This problem is considered by Theil (1971). Equation (4.2.10) is of importance in the study of empirical Bayes estimators (see Gruber (1979)).

4.2.5. Remarks

1. The condition that $V'FU = 0$ restricts the form of the prior dispersion as explained below. Since $UU' + VV' = I$,

$$F = (UU' + VV')F(UU' + VV')$$

$$= [U \ V] \begin{bmatrix} U' \\ V' \end{bmatrix} F[U \ V] \begin{bmatrix} U' \\ V' \end{bmatrix}$$

$$= [U \ \ V] \begin{bmatrix} U'FU & U'FV \\ V'FU & V'FV \end{bmatrix} \begin{bmatrix} U' \\ V' \end{bmatrix} \qquad (4.2.11)$$

$$= [U \ \ V] \begin{bmatrix} M_1 & 0 \\ 0 & M_2 \end{bmatrix} \begin{bmatrix} U' \\ V' \end{bmatrix} = UM_1U' + VM_2V',$$

where $M_1 = U'FU$ and $M_2 = V'FV$.

2. Assume that either $p'\beta$ is estimable or $V'FU = 0$. The condition that $U'FU$ is positive definite is necessary and sufficient for equivalence of (3.5.7) and (4.2.4). The proof is given below.

Suppose $t = \text{rank } (U'FU) \le s$. Then there is an $s \times t$ matrix C such that $U'FU = CC'$. From (3.5.7),

$$p'\hat{\beta} = p'\theta + p'UCC'U'X'(XUCC'U'X' + \sigma^2 I)^{-1}(Y - X\theta). \quad (4.2.12)$$

From (4.1.7)

$$C'U'X'(XUCC'U'X' + \sigma^2 I)^{-1}$$

$$= \frac{1}{\sigma^2} C'U'X' - \frac{1}{\sigma^2} C'U'X'XUC[\frac{C'U'X'XUC}{\sigma^2} + I]^{-1}\frac{C'U'X'}{\sigma^2}$$

$$= [C'U'X'XUC + \sigma^2 I]^{-1}C'U'X'. \qquad (4.2.13)$$

From (4.2.12) and (4.2.13),

$$p'\hat{\beta} = p'\theta + p'UC[C'U'X'XUC + \sigma^2 I]^{-1}C'U'X'(Y - X\theta). \quad (4.2.14)$$

When $U'FU$ is positive definite, C is $s \times s$ and is nonsingular. From (4.2.14)

$$p'\hat{\beta} = p'\theta + p'U[U'X'XU + \sigma^2(CC')^{-1}]^{-1}U'X'(Y - X\theta)$$

$$= p'\theta + p'[UU'X'XUU' + \sigma^2(UU'FUU')^+]^+X'(Y - X\theta)$$

$$= p'\theta + p'[X'X + \sigma^2(UU'FUU')^+]^+X'(Y - X\theta). \qquad (4.2.15)$$

Suppose rank $(U'FU) = t < s$. The singular value decomposition (SVD) of C is

$$C = [A' \quad B'] \begin{bmatrix} \Delta^{1/2} & 0 \\ 0 & 0 \end{bmatrix} \begin{bmatrix} P' \\ Q' \end{bmatrix}$$

$$\qquad (4.2.16)$$

$$= A'\Delta^{1/2}P'.$$

Let

$$M = \begin{bmatrix} \Delta^{1/2} & 0 \\ 0 & 0 \end{bmatrix} \begin{bmatrix} A \\ B \end{bmatrix} \Lambda[A' \quad B'] \begin{bmatrix} \Delta^{1/2} & 0 \\ 0 & 0 \end{bmatrix}.$$

Then

$$[C'\Lambda C + \sigma^2 I]^{-1}$$

$$= \left[[P \; Q]M \begin{bmatrix} P' \\ Q' \end{bmatrix} + [P \; Q] \begin{bmatrix} P' \\ Q' \end{bmatrix} \sigma^2 \right]^{-1}$$

$$= [P \; Q] \begin{bmatrix} \Delta^{1/2}A\Lambda A'\Delta^{1/2} + \sigma^2 I & 0 \\ \\ 0 & \sigma^2 I \end{bmatrix}^{-1} \begin{bmatrix} P' \\ Q' \end{bmatrix} \qquad (4.2.17)$$

$$= P(\Delta^{1/2}A\Lambda A'\Delta^{1/2} + \sigma^2 I)^{-1}P' + \frac{1}{\sigma^2}QQ'.$$

Then

$$C[C'\Lambda C + \sigma^2 I]^{-1}C'$$

$$= A'\Delta^{1/2}(\Delta^{1/2}A\Lambda A'\Delta^{1/2} + \sigma^2 I)^{-1}\Delta^{1/2}A \qquad (4.2.18)$$

$$= A'(A\Lambda A' + \sigma^2\Delta^{-1})^{-1}A$$

and

$$UC[C'\Lambda C + \sigma^2 I]^{-1}C'U'$$

$$= [U A'AU X'XU A'AU' + \sigma^2 U A'\Delta^{-1}AU']^+ \qquad (4.2.19)$$

$$= [U A'A\Lambda A'AU' + \sigma^2(UU'FUU')^+]^+ = H.$$

Now $U A' A U' X' X U A' A U' = X'X$ iff

$$A' A \Lambda A' A = \Lambda. \tag{4.2.20}$$

But Λ is of rank s so $A'A$ is of full rank s. Then $B = 0$ and $U'FU$ is of full rank. Hence $U'FU$ is positive definite if (4.2.15) holds. Otherwise

$$p'\hat{\beta} = p'\theta + p'HX'(Y - X\theta) \tag{4.2.21}$$

It is a worthwhile observation that Equations (4.2.7) and (4.2.10) may be derived without assuming that $U'FU$ is positive definite. To see this, notice that, from the SVD of X,

$$XUU' = [S'\ T']\begin{bmatrix} \Lambda^{1/2} & 0 \\ 0 & 0 \end{bmatrix}\begin{bmatrix} U' \\ V' \end{bmatrix} UU'$$

$$= [S'\ T']\begin{bmatrix} \Lambda^{1/2} & 0 \\ 0 & 0 \end{bmatrix}\begin{bmatrix} U' \\ 0 \end{bmatrix} \tag{4.2.22}$$

$$= [S'\ T']\begin{bmatrix} \Lambda^{1/2}U' \\ 0 \end{bmatrix}.$$

Consequently

$$UU'X' = [U\Lambda^{1/2}\ 0]\begin{bmatrix} S \\ T \end{bmatrix}. \tag{4.2.23}$$

From (4.2.23)

$$XUU'FUU'X' = [S'\ T']\begin{bmatrix} \Lambda^{1/2}U'FU\Lambda^{1/2} & 0 \\ 0 & 0 \end{bmatrix}\begin{bmatrix} S \\ T \end{bmatrix}. \tag{4.2.24}$$

Let $D = \Lambda^{1/2} U' F U \Lambda^{1/2} + \sigma^2 I$. Since $S'S + T'T = I$, from (4.2.24),

$$(XUU'FUU'X' + \sigma^2 I)^{-1} = \left\{ [S'\ T'] \begin{bmatrix} D & 0 \\ 0 & \sigma^2 I \end{bmatrix} \begin{bmatrix} S \\ T \end{bmatrix} \right\}^{-1}$$

$$= \begin{bmatrix} S \\ T \end{bmatrix}^{-1} \begin{bmatrix} D & 0 \\ 0 & \sigma^2 I \end{bmatrix}^{-1} [S'\ T']^{-1}$$

$$= [S'\ T'] \begin{bmatrix} D^{-1} & 0 \\ 0 & \frac{1}{\sigma^2} I \end{bmatrix} \begin{bmatrix} S \\ T \end{bmatrix}.$$

$$(4.2.25)$$

Also

$$U'X' = U'[U\ V] \begin{bmatrix} \Lambda^{1/2} & 0 \\ 0 & 0 \end{bmatrix} \begin{bmatrix} S \\ T \end{bmatrix}$$

$$(4.2.26)$$

$$= [I\ 0] \begin{bmatrix} \Lambda^{1/2} & 0 \\ 0 & 0 \end{bmatrix} \begin{bmatrix} S \\ T \end{bmatrix}.$$

From (4.2.25) and (4.2.26)

$$UU'X'(XUU'FUU'X' + \sigma^2 I)^{-1}$$

$$= U[I\ 0] \begin{bmatrix} \Lambda^{1/2} & 0 \\ 0 & 0 \end{bmatrix} \begin{bmatrix} S \\ T \end{bmatrix} [S'\ T'] \begin{bmatrix} D^{-1} & 0 \\ 0 & 0 \end{bmatrix} \begin{bmatrix} S \\ T \end{bmatrix}$$

$$(4.2.27)$$

$$= U\Lambda^{1/2} (\Lambda^{1/2} U' F U \Lambda^{1/2} + \sigma^2 I)^{-1} S$$

$$= U(U'FU + \sigma^2 \Lambda^{-1})^{-1} \Lambda^{-1/2} S.$$

Then

$$p'\hat{\beta} = p'\theta + p'FU(U'FU + \sigma^2\Lambda^{-1})^{-1}\Lambda^{-1/2}S(Y - X\theta). \quad (4.2.28)$$

Now

$$\Lambda^{-1/2}SX = \Lambda^{-1/2}SS'\Lambda^{1/2}U' = U'. \quad (4.2.29)$$

Thus, from (4.2.28) and (4.2.29),

$p'\hat{\beta}$

$$= p'\theta + p'FU(U'FU + \sigma^2\Lambda^{-1})^{-1}[\Lambda^{-1/2}SY - U'\theta]$$

$$= p'\theta + p'FU(U'FU + \sigma^2\Lambda^{-1})^{-1}[U'U\Lambda^{-1}U'U\Lambda^{1/2}SY - U'\theta]$$

$$= p'\theta + p'F(UU'FUU' + \sigma^2(X'X)^+)^+[(X'X)^+X'Y - \theta]$$

$$= p'\theta + p'F[UU'FUU' + \sigma^2(X'X)^+]^+[b - \theta].$$
$$(4.2.30)$$

Equation (4.2.30) and (4.2.7) are the same.

Example 4.2.1. Conditions of Validity of Alternative Forms of the BE Illustrated. Consider the model

$$\begin{bmatrix} Y_{11} \\ Y_{12} \\ Y_{21} \\ Y_{22} \end{bmatrix} = \begin{bmatrix} 1 & 1 & 0 \\ 1 & 1 & 0 \\ 1 & 0 & 1 \\ 1 & 0 & 1 \end{bmatrix} \begin{bmatrix} \mu \\ \alpha_1 \\ \alpha_2 \end{bmatrix} + \epsilon,$$

where $E(\epsilon) = 0$, $D(\epsilon) = \sigma^2 I$,

$$E \begin{bmatrix} \mu \\ \alpha_1 \\ \alpha_2 \end{bmatrix} = \theta, \quad D \begin{bmatrix} \mu \\ \alpha_1 \\ \alpha_2 \end{bmatrix} = F.$$

The reader can show that the estimable parametric functions are linear combinations of $\mu + \alpha_1$ and $\mu + \alpha_2$. Thus $\alpha_1 - \alpha_2$ is estimable but α_1 is not. When

$$F = cI$$

with c a scalar the alternative forms of both $\hat{\alpha}_1$ and $\hat{\alpha}_1 - \hat{\alpha}_2$ may be written (Exercise 4.2.1). But for the prior dispersion

$$F = \begin{bmatrix} 3 & 0 & 0 \\ 0 & 2 & 0 \\ 0 & 0 & 1 \end{bmatrix}$$

alternative forms may be obtained for $\hat{\alpha}_1 - \hat{\alpha}_2$ but not for $\hat{\alpha}_1$.

$$V'FU = \begin{bmatrix} \frac{1}{\sqrt{3}} & -\frac{1}{\sqrt{3}} & -\frac{1}{\sqrt{3}} \end{bmatrix} \begin{bmatrix} 3 & 0 & 0 \\ 0 & 2 & 0 \\ 0 & 0 & 1 \end{bmatrix} \begin{bmatrix} \frac{2}{\sqrt{6}} & 0 \\ \frac{1}{\sqrt{6}} & \frac{1}{\sqrt{2}} \\ \frac{1}{\sqrt{6}} & -\frac{1}{\sqrt{2}} \end{bmatrix}$$

$$= \begin{bmatrix} \frac{3}{\sqrt{18}} & -\frac{1}{\sqrt{6}} \end{bmatrix} \neq 0.$$

(The reader should obtain the SVD of the $X'X$ matrix, Exercise 4.2.1.)

Exercise 4.2.1. For Example 4.2.1 above

A. Characterize the estimable parametric functions.
B. Find the singular value decomposition of $X'X$.

C. Show that $U'FU$ is PD for both priors.

D. For $F = cI$ show that $V'FU = 0$.

E. Exhibit those alternative forms possible for $\hat{\alpha}_1$ and $\hat{\alpha}_1 - \hat{\alpha}_2$ for both above listed priors.

Exercise 4.2.2.

For each of the following models find the alternative forms of the BE for the given priors and estimable parametric functions.

A.

1. Model

$$Y = \begin{bmatrix} 1_3 & 1_3 & 0 & 0 \\ 1_3 & 0 & 1_3 & 0 \\ 1_3 & 0 & 0 & 1_3 \end{bmatrix} \begin{bmatrix} \mu \\ \alpha_1 \\ \alpha_2 \\ \alpha_3 \end{bmatrix} + \epsilon.$$

2. Assumptions

$$E(\epsilon|\beta) = 0, \qquad D(\epsilon|\beta) = I,$$

$$E(\beta) = \begin{bmatrix} 1 \\ -2 \\ 4 \\ 0 \end{bmatrix}, \qquad D(\beta) = \begin{bmatrix} 3 & -1 & -1 & -1 \\ -1 & \frac{5}{3} & -\frac{2}{3} & -\frac{1}{3} \\ -1 & -\frac{2}{3} & \frac{5}{3} & -\frac{1}{3} \\ -1 & -\frac{1}{3} & -\frac{1}{3} & \frac{5}{3} \end{bmatrix}.$$

3. Estimable Parametric Function

$$\alpha_1 + \alpha_2 - 2\alpha_3.$$

B.

1. Model

$$\begin{bmatrix} Y_1 \\ Y_2 \end{bmatrix} = \begin{bmatrix} 1 & 1 & -2 \\ -1 & 1 & 0 \end{bmatrix} \begin{bmatrix} \beta_1 \\ \beta_2 \\ \beta_3 \end{bmatrix} + \begin{bmatrix} \epsilon_1 \\ \epsilon_2 \\ \epsilon_3 \end{bmatrix},$$

$$E(\epsilon|\beta) = 0, \qquad D(\epsilon|\beta) = \sigma^2 I.$$

2. Assumptions

$$E(\beta) = \begin{bmatrix} 1 \\ 1 \\ -2 \\ 0 \end{bmatrix}, \quad D(\beta) = \begin{bmatrix} 4 & 0 & 0 \\ 0 & 2 & 0 \\ 0 & 0 & 0 \end{bmatrix}.$$

3. Estimable Parametric Function

$$3\beta_1 + \beta_2 - 2\beta_3.$$

4.3 The Equivalence of the Generalized Ridge Estimator and the Bayes Estimator

In Section 3.2 the form of the generalized ridge estimator was obtained (see (3.2.4)) by generalizing the frequentist argument of Hoerl and Kennard (1970). Using the alternative form (4.2.6) it will be shown that for the class of prior dispersions in equations (4.3.4) below the generalized ridge estimator is a special case of the Bayes estimator.
  Since

$$(X'X_{/}(X'X)^+ X' = U\Lambda U'U\Lambda^{-1}U'X' = UU'X' = X', \quad (4.3.1)$$

Equation (3.2.4) may be rewritten

$$p'\hat{\beta}gr = p'[H + \lambda X'X]^+[H\theta + \lambda X'Y]. \quad (4.3.2)$$

The form of the prior dispersion that (4.3.2) is Bayes with respect to must be specified. From (3.2.6) the prior dispersion F should be such that $U'FU$ is positive definite. Since

$$F = (UU' + VV')F(UU' + VV')$$

$$= [U \ V] \begin{bmatrix} U'FU & U'FV \\ U'FU & V'FV \end{bmatrix} \begin{bmatrix} U' \\ V' \end{bmatrix}, \quad (4.3.3)$$

the matrix F may be written in the form

$$F = [U \ V] \begin{bmatrix} A^{-1} & C_1 \\ C_1' & C_2 \end{bmatrix} \begin{bmatrix} U' \\ V' \end{bmatrix} \lambda \sigma^2, \qquad (4.3.4)$$

where A is positive definite. Then

$$U'FU = U'[U \ V] \begin{bmatrix} A^{-1} & C_1 \\ C_1' & C_2 \end{bmatrix} \begin{bmatrix} U' \\ V' \end{bmatrix} U \lambda \sigma^2$$

$$(4.3.5)$$

$$= [I \ 0] \begin{bmatrix} A^{-1} & C_1 \\ C_1' & C_2 \end{bmatrix} \begin{bmatrix} I \\ 0 \end{bmatrix} \lambda \sigma^2 = A^{-1} \lambda \sigma^2.$$

Then

$$A = (U'FU)^{-1} \lambda \sigma^2. \qquad (4.3.6)$$

Thus, if

$$H = U(U'FU)^{-1}U'\lambda \sigma^2 = (UU'FUU')^{+}\lambda \sigma^2, \qquad (4.3.7)$$

substitutions of (4.3.7) into (4.3.2) yields the alternative form of the Bayes estimator (4.2.6).

Thus, for a prior dispersion of the form (4.3.4) the generalized ridge estimator is a special case of the Bayes estimator. This is also true when

$$H = (UU'FUU')^{+}\lambda \sigma^2 + VTU'.$$

If $(U'FU)$ is PD and

$$H = (UU'FUU')^{+}\lambda \sigma^2,$$

the Bayes estimator is a generalized ridge estimator.

Example 4.3.1. Ridge Regression Estimator. Consider the linear model Example 3.1.1. Let

$$E(\beta) = 0 \quad \text{and} \quad D(\beta) = \begin{bmatrix} \frac{1}{k} & 0 & 0 \\ 0 & 0 & 0 \\ 0 & 0 & \frac{1}{k} \end{bmatrix}.$$

Then since

$$UU' = \begin{bmatrix} \frac{1}{\sqrt{2}} & \frac{1}{\sqrt{2}} \\ 0 & 0 \\ \frac{1}{\sqrt{2}} & -\frac{1}{\sqrt{2}} \end{bmatrix} \begin{bmatrix} \frac{1}{\sqrt{2}} & 0 & \frac{1}{\sqrt{2}} \\ \frac{1}{\sqrt{2}} & 0 & -\frac{1}{\sqrt{2}} \end{bmatrix} = \begin{bmatrix} 1 & 0 & 0 \\ 0 & 0 & 0 \\ 0 & 0 & 1 \end{bmatrix},$$

$$UU'FUU' = \begin{bmatrix} \frac{1}{k} & 0 & 0 \\ 0 & 0 & 0 \\ 0 & 0 & \frac{1}{k} \end{bmatrix}.$$

Thus,

$$H = \begin{bmatrix} k & 0 & 0 \\ 0 & 0 & 0 \\ 0 & 0 & k \end{bmatrix} \lambda \sigma^2.$$

Let $d = k\lambda\sigma^2$. The ridge estimator is

$$\begin{bmatrix} \hat{\beta}_0 \\ \hat{\beta}_1 \\ \hat{\beta}_2 \end{bmatrix} = \begin{bmatrix} \frac{3}{2}+d & \frac{1}{2} & 0 \\ \frac{1}{2} & \frac{1}{2} & 0 \\ 0 & 0 & 1+d \end{bmatrix}^{-1} \begin{bmatrix} Y_1 + \frac{1}{\sqrt{2}}\cdot Y_2 \\ Y_1 + \frac{1}{\sqrt{2}}\cdot Y_3 \\ Y_3 \end{bmatrix}$$

$$= \begin{bmatrix} 0 \\ 2(Y_1 + \frac{1}{\sqrt{2}}\frac{1}{2}) \\ \frac{1}{1+d}Y_3 \end{bmatrix}.$$

The reader can verify that $U'FU$ is PD; see Exercise 4.3.2.

Exercise 4.3.1. Let

$$Y = \begin{bmatrix} 1 & 0 & 1 & 0 \\ 1 & 0 & 0 & 1 \\ 0 & 1 & 1 & 0 \\ 0 & 1 & 0 & 1 \end{bmatrix} \begin{bmatrix} \alpha_1 \\ \alpha_2 \\ \beta_1 \\ \beta_2 \end{bmatrix} + \begin{bmatrix} \epsilon_1 \\ \epsilon_2 \\ \epsilon_3 \\ \epsilon_4 \end{bmatrix}.$$

For generalized ridge regression estimator (4.3.2) with $\theta = 0$ find H so that the estimator is Bayes with respect to a prior distribution with dispersion

A.

$$F = I_4 + J_{4\times4}.$$

B.

$$F = \begin{bmatrix} 2 & 1 & 1 & 0 \\ 1 & 2 & 1 & 0 \\ 1 & 1 & 2 & 0 \\ 0 & 0 & 0 & 0 \end{bmatrix}.$$

Exercise 4.3.2. Verify that in Example 4.3.1 $U'FU$ is PD.

4.4 The Equivalence of the Mixed Estimator and the Bayes Estimator

In this section the mixed estimator of Section 3.3 and the Bayes estimator of Section 3.5 is considered. When $U'FU$ is positive definite and when $R = A_1 U'$:

1. the equivalence of the Bayes estimator and the mixed estimator will be shown;
2. it will be shown that sample and prior information may be exchanged;
3. alternative forms of the mixed estimator similar to the alternative forms of the BE described in Section 3.2 and their interpretation will be given.

Rewrite the model (3.3.1) as two separate models

$$Y = X\beta + \epsilon \qquad \qquad \text{Model I} \qquad \qquad (4.4.1)$$

with

$$E(\epsilon) = 0 \quad \text{and} \quad D(\epsilon) = \sigma^2 I \qquad \qquad (4.4.2)$$

and

$$r = R\beta + \phi \qquad \qquad \text{Model II} \qquad \qquad (4.4.3)$$

with

$$E(\phi) = 0 \quad \text{and} \quad D(\phi) = \tau^2 I. \qquad \qquad (4.4.4)$$

Equation (4.4.1) will be referred to as model I and (4.4.3) as model II.

Recall that $p'\beta$ is estimable with respect to model I (X estimable) if $p' = d'U'$.

The SVD of

$$R = [A' \ B'] \begin{bmatrix} \Delta^{1/2} & 0 \\ 0 & 0 \end{bmatrix} \begin{bmatrix} P' \\ Q' \end{bmatrix} = A'\Delta^{1/2}P'. \qquad (4.4.5)$$

Now $p'\beta$ is estimable with respect to model II (R estimable) iff $p' = \ell'P'$ for some ℓ; in other words p is in the column space of P.

From (3.3.5) the mixed estimator takes the form

$$p'\widehat{\beta}_m = p'(\tau^2 X'X + \sigma^2 R'R)^+(\tau^2 X'Y + \sigma^2 R'r). \qquad (4.4.6)$$

Observe that, since $(X'X)^+(X'X)X' = X'$ and $(R'R)^+(R'R)R' = R'$,

$$p'\widehat{\beta}_m = p'(\tau^2 X'X + \sigma^2 R'R)^+[X'Xb_1\tau^2 + R'Rb_2\sigma^2], \qquad (4.4.7)$$

where $b_1 = (X'X)^+X'Y$ and $b_2 = (R'R)^+R'r$ are the least square estimators for model I and model II respectively.

It can be verified (Exercise 4.4.5) that if $p'\beta$ is X estimable, i.e. $p' = d'U'$ and $R = A'U'$, then equation (4.4.7) is algebraically equivalent to

$$p'\widehat{\beta} = p'b_2 + p'(R'R)^+X'[X(R'R)^+X'\tau^2 + \sigma^2 I]^{-1}(Y - Xb_2)\tau^2. \qquad (4.4.8)$$

Observe that (4.4.8) is algebraically the same as the BE with b_2 replacing θ and $\sigma^2(R'R)^+$ replacing F.

Likewise, if $p'\beta$ is R estimable and $P'X'XP$ is PD it can be shown that (4.4.7) is algebraically equivalent to

$$p'\widehat{\beta} = p'b_1 + p'(X'X)^+R'[\sigma^2 R(X'X)^+R' + \tau^2 I]^{-1}(Y - Rb_1)\sigma^2. \qquad (4.4.9)$$

Equation (4.4.9) is algebraically the same as the BE (4.4.8) with the roles of R and X exchanged, b_1 in place of θ and $(X'X)^+\sigma^2$ in place of X.

When $U'(R'R)^+U$ and $P'(X'X)^+P$ are positive definite the sample and the prior information can be exchanged and equivalent estimators will result.

From (4.4.8) the following alternative forms of the mixed estimators are valid for model I. If $p'\beta$ is estimable

$$p'\widehat{\beta} = p'b_2 + p'\left[\frac{X'X}{\sigma^2} + \frac{R'R}{\tau^2}\right]^+ \frac{X'}{\sigma^2}(Y - Xb_2) \qquad (4.4.10)$$

$$= p'\left[\frac{X'X}{\tau^2} + \frac{R'R}{\sigma^2}\right]^+ \left[\frac{X'X}{\sigma^2}b_1 + \frac{R'R}{\tau^2}b_2\right]. \qquad (4.4.11)$$

Also,

$$p'\widehat{\beta} = p'b_2 + p'(R'R)^+[(X'X)^+\sigma^2 + (R'R)^+\tau^2]^+(b_1 - b_2)\tau^2$$

$$= p'(X'X)^+[\tau^2(R'R)^+ + \sigma^2(X'X)^+]^+b_2\sigma^2 \qquad (4.4.12)$$

$$+ p'(R'R)^+[\tau^2(R'R)^+ + \sigma^2(X'X)^+]^+b_1\tau^2.$$

For (4.4.1) to hold true $p'\widehat{\beta}$ need not be estimable. Also

$$p'\widehat{\beta} = p'b_1 - p'(X'X)^+[\tau^2(R'R)^+ + \sigma^2(X'X)^+]^+(b_1 - b_2)\sigma^2. \qquad (4.4.13)$$

The exchangeability of the sample and the prior information may be used to obtain the corresponding alternative forms for model II. Simply exchange X and R, b_1 and b_2, σ^2 and τ^2 in Equations (4.4.10)–(4.4.13).

Thus for model II, for estimable parametric functions,

$$p'\widehat{\beta} = p'b_1 + p'\left[\frac{X'X}{\sigma^2} + \frac{R'R}{\tau^2}\right]^+ \frac{R'}{\tau^2}(r - Rb_1)$$

$$= p'\left[\frac{X'X}{\sigma^2} + \frac{R'R}{\tau^2}\right]^+ \left[\frac{X'X}{\sigma^2}b_1 + \frac{R'R}{\tau^2}b_2\right]$$

$$= p'b_2 - p'(R'R)^+\left[\tau^2(R'R)^+ + \sigma^2(X'X)^+\right]^+(b_2 - b_1)\sigma^2. \qquad (4.4.14)$$

The alternative form in (4.4.11) is the same for both models.
For the argumented model

$$\begin{bmatrix} Y \\ r \end{bmatrix} = \begin{bmatrix} X \\ R \end{bmatrix} \beta + \begin{bmatrix} \epsilon \\ \eta \end{bmatrix} \tag{4.4.15}$$

assume

$$E(\beta) = \theta, \qquad D(\beta) = F, \tag{4.4.16}$$

where θ is an unknown vector and F is a known non-negative definite matrix. Also, as before

$$E(\epsilon|\beta) = 0, \qquad D(\epsilon|\beta) = \sigma^2 I \tag{4.4.17}$$

and

$$E(\eta|\beta) = 0, \qquad D(\eta|\beta) = \tau^2 I.$$

If there is a t where

$$p' = t' \begin{bmatrix} X \\ R \end{bmatrix}, \tag{4.4.18}$$

i.e. $p'\beta$ is X estimable, then the solution to the optimization problem in Section 3.5 yields the mixed estimator.

Thus, the mixed estimator may be thought of in three ways:

1. as a weighted least square estimator;
2. as equivalent to a linear BE with known prior mean;
3. as an optimum estimator with unknown prior mean if $p'\beta$ is estimable.

Notice that to get the forms of the BE for model II from those of model I simply interchange the sample and prior information. Thus, under the conditions mentioned above the sample and prior information are interchangeable. Another way to view prior information is as additional observations, i.e., more sampling.

An important question suggests itself. To what extent are sample and prior information interchangeable under more general conditions? This is taken up in Section 4.6.

However, the case where $U'FU$ is positive definite does include all of the standard examples of ridge estimators discussed in the literature. This shall be taken up first.

Of course, if X and R are of full rank, then sample and prior information are interchangeable. However consider Example 4.4.1 below.

Example 4.4.1. Exchange of Sample and Prior Information.
Let

$$
Y = \begin{bmatrix} I_3 & I_3 & 0 & 0 \\ I_3 & 0 & I_3 & 0 \\ I_3 & 0 & 0 & I_3 \end{bmatrix} \begin{bmatrix} \mu \\ \tau_1 \\ \tau_2 \\ \tau_3 \end{bmatrix} + \epsilon \qquad \text{Model I}
$$

and

$$
r = \begin{bmatrix} \frac{1}{2} & \frac{1}{2} & \frac{1}{2} & \frac{1}{2} \\ \frac{1}{\sqrt{2}} & -\frac{1}{\sqrt{2}} & 0 & 0 \\ \frac{1}{\sqrt{6}} & \frac{1}{\sqrt{6}} & -\frac{2}{\sqrt{6}} & 0 \end{bmatrix} \begin{bmatrix} \mu \\ \tau_1 \\ \tau_2 \\ \tau_3 \end{bmatrix} + \phi. \qquad \text{Model II}
$$

Then the SVD of $X'X$ is (as the reader may verify in Exercise 4.4.1)

$$
X'X = U\Lambda U'
$$

with

$$U = \begin{bmatrix} \frac{3}{\sqrt{12}} & 0 & 0 \\[2mm] \frac{1}{\sqrt{12}} & 0 & -\frac{2}{\sqrt{6}} \\[2mm] \frac{1}{\sqrt{12}} & -\frac{1}{\sqrt{12}} & \frac{1}{\sqrt{6}} \\[2mm] \frac{1}{\sqrt{12}} & \frac{1}{\sqrt{12}} & \frac{1}{\sqrt{6}} \end{bmatrix}$$

and

$$\Lambda = \begin{bmatrix} 12 & 0 & 0 \\ 0 & 3 & 0 \\ 0 & 0 & 3 \end{bmatrix}.$$

The reader may also verify that the SVD of $R'R$ is

$$R'R = \begin{bmatrix} \frac{1}{2} & \frac{1}{\sqrt{2}} & \frac{1}{\sqrt{6}} \\[2mm] \frac{1}{2} & -\frac{1}{\sqrt{2}} & \frac{1}{\sqrt{6}} \\[2mm] \frac{1}{2} & 0 & -\frac{2}{\sqrt{6}} \\[2mm] \frac{1}{2} & 0 & 0 \end{bmatrix} \begin{bmatrix} 4 & 0 & 0 \\ 0 & 2 & 0 \\ 0 & 0 & 1 \end{bmatrix} \begin{bmatrix} \frac{1}{2} & \frac{1}{2} & \frac{1}{2} & \frac{1}{2} \\[2mm] \frac{1}{\sqrt{2}} & -\frac{1}{\sqrt{2}} & 0 & 0 \\[2mm] \frac{1}{\sqrt{6}} & \frac{1}{\sqrt{6}} & -\frac{2}{\sqrt{6}} & 0 \end{bmatrix}.$$

Now

$$(X'X)^+ = \frac{1}{144} \begin{bmatrix} 9 & 3 & 3 & 3 \\ 3 & 33 & -15 & -15 \\ 3 & -15 & 33 & -15 \\ 3 & -15 & -15 & 33 \end{bmatrix},$$

$$(R'R)^+ = \frac{1}{48} \begin{bmatrix} 23 & -1 & -13 & 3 \\ -1 & 23 & -13 & 3 \\ -13 & -13 & 35 & 3 \\ 3 & 3 & 3 & 3 \end{bmatrix},$$

$$P'(X'X)^+ P = \frac{1}{144} \begin{bmatrix} 9 & \frac{6}{\sqrt{2}} & \frac{6}{\sqrt{6}} \\ \frac{6}{\sqrt{2}} & \frac{36}{\sqrt{2}} & -\frac{60}{\sqrt{12}} \\ \frac{6}{\sqrt{6}} & -\frac{60}{\sqrt{12}} & 31 \end{bmatrix},$$

and

$$U'(R'R)^+ U = \frac{1}{48} \begin{bmatrix} \frac{47}{3} & \frac{32}{\sqrt{24}} & -\frac{16}{\sqrt{72}} \\ \frac{32}{\sqrt{24}} & 16 & -\frac{64}{\sqrt{12}} \\ -\frac{16}{\sqrt{72}} & -\frac{64}{\sqrt{12}} & \frac{88}{3} \end{bmatrix}.$$

Both $P'(X'X)^+ P$ and $U'(R'R)^+ U$ are clearly positive definite as the reader may show. Thus, sample and prior information may be interchanged.

With model II as design matrix the prior assumptions are for the BE

$$E(\beta) = \begin{bmatrix} \frac{18}{144}[9Y_{..}] \\ \frac{1}{144}(3Y_{..} + 33Y_{1.} - 15Y_{2.} - 15Y_{3.}) \\ \frac{1}{144}(3Y_{..} - 15Y_{1.} + 33Y_{2.} - 15Y_{3.}) \\ \frac{1}{144}(3Y_{..} - 15Y_{1.} - 15Y_{3.} + 33Y_{3.}) \end{bmatrix}$$

and

$$D(\beta) = \frac{\sigma^2}{144} \begin{bmatrix} 9 & 3 & 3 & 3 \\ 3 & 33 & -15 & -15 \\ 3 & -15 & 33 & -15 \\ 3 & -15 & -15 & 33 \end{bmatrix}.$$

With model I as design matrix the reader should write down the form of the prior mean and dispersion.

Exercise 4.4.1. For Example 4.4.1 above
A. Verify the SVD of $X'X$ and $R'R$.
B. Verify the Moore Penrose inverses of $(X'X)^+$ and $(R'R)^+$.
C. Show that $P'(X'X)^+P$ and $U'(R'R)^+U$ are positive definite.

Exercise 4.4.2. Consider the data below for the augmented linear model (4.4.1) and (4.4.2):

$$Y' = (0.43 \ 0.31 \ 0.32 \ 0.46 \ 1.25 \ 0.44 \ 0.52 \ 0.29 \ 1.29 \ 0.35),$$

$$X = \begin{bmatrix} 1 & 2.1 & 3 \\ 1 & 1.1 & 4 \\ 1 & 0.9 & 5 \\ 1 & 1.6 & 4 \\ 1 & 6.2 & 4 \\ 1 & 2.3 & 3 \\ 1 & 1.8 & 6 \\ 1 & 1.0 & 5 \\ 1 & 8.9 & 3 \\ 1 & 2.4 & 2 \end{bmatrix}, \quad R = \begin{bmatrix} 1 & 1.2 & 4 \\ 1 & 4.7 & 3 \\ 1 & 3.5 & 2 \\ 1 & 2.9 & 3 \\ 1 & 1.4 & 4 \end{bmatrix},$$

$$r' = (0.35 \ 0.78 \ 0.43 \ 0.47 \ 0.38).$$

A. Find b_1 and b_2.
B. Find the mixed estimator. Assume $\epsilon \sim N(0, \sigma^2 I)$ and $\phi \sim N(0, \tau^2 I)$. Estimate σ^2 and τ^2 from b_1 and b_2.
C. Calculate the statistic

$$\hat{\gamma} = (r - Rb)'[\hat{\tau}^2 I + \hat{\sigma}^2 R(X'X)^{-1}R']^{-1}(r - Rb)$$

and compare it to $\chi^2_{5,0.5}$. If $\hat{\gamma}$ is too large the sample and prior information may be incompatible. (This test was developed by H. Theil(1963).)

Exercise 4.4.3. Consider the linear model

$$Y_1 = \sqrt{2}(\beta_1 + \beta_3) + \epsilon_1,$$

$$Y_2 = 2\beta_2 + \epsilon_2$$

and

$$r_1 = \beta_1 + \beta_2 + \beta_3 + \phi_1,$$

$$r_2 = \beta_1 + 6\beta_2 + \beta_3 + \phi_2$$

with

$$E(\epsilon) = 0, \qquad D(\epsilon) = \sigma^2 I,$$

$$E(\phi) = 0, \qquad D(\phi) = \tau^2 I.$$

Find the mixed estimator and its alternative forms if possible.

Exercise 4.4.4. Same problem as in Exercise 4.4.3 with

$$r_1 = \sqrt{\frac{k_1}{2}}(\beta_1 + \beta_3) + \epsilon_1, \qquad k_1, k_2 > 0,$$

$$r_2 = \sqrt{k_2}\beta_2 + \epsilon_2.$$

Exercise 4.4.5. Show that Equations (4.4.8) and (4.4.9) are both algebraically equivalent to (4.4.7) under the conditions given in the text.

4.5 Ridge Estimators in the Literature as Special Cases of the BE, Minimax Estimators, or Mixed Estimators

Each of the ridge type estimators described in the statistical litera-ture can be shown to be a special case of the Bayes estimator and the minimax estimators.

To do this let $\theta = 0$ and $H = \sigma^2(UU'FUU')^+$ in (4.2.6). Then rewrite (4.2.6) in the form

$$p'\hat{\beta} = p'[X'X + H]^+ X'Y. \qquad (4.5.1)$$

This is the generalized ridge estimator. In the full rank case with F PD, C.R. Rao (1975) observed that, when $\theta = 0$ and $G = \sigma^2 F^{-1}$,

$$p'\hat{\beta} = p'(X'X + G)^{-1} X'Y \qquad (4.5.2)$$

was the ridge estimator of the Hoerl and Kennard (1970).

Equation (4.5.1) will now be rewritten in another form so that the analogous result for the non-full rank case can be obtained.

Observe that, for (4.2.6) with $\theta = 0$, by (2.1.21)

$$p'\hat{\beta} = p'[X'X + \sigma^2(UU'FUU')^+]^+ X'Y$$

$$\qquad (4.5.3)$$

$$= p'U[\Lambda + \sigma^2(U'FU)^{-1}]^{-1}U'X'Y.$$

Let T be any $(n - s) \times (n - s)$ NND matrix. Then, from (2.1.8) and

(2.1.23),

$$p'\hat{\beta} = p'[U(\Lambda + \sigma^2(U'FU)^{-1})^{-1}U' + VTV']X'Y$$

$$= p'[U \ V]\begin{bmatrix} (\Lambda + (U'FU)^{-1}\sigma^2)^{-1} & 0 \\ 0 & T \end{bmatrix}\begin{bmatrix} U' \\ V' \end{bmatrix}X'Y$$

$$\text{(4.5.4)}$$

$$= p'[U \ V]\begin{bmatrix} \Lambda + \sigma^2(U'FU)^{-1} & 0 \\ 0 & T^+ \end{bmatrix}^+\begin{bmatrix} U' \\ V' \end{bmatrix}X'Y$$

$$= p'[X'X + \sigma^2(UU'FUU')^+ + VT^+V']^+X'Y.$$

Thus, let G be of the form $H + VT^+V'$. Then

$$p'\hat{\beta} = p'[X'X + G]^+X'Y, \qquad (4.5.5)$$

where G is a matrix of rank between s and m inclusive.

When $G = kI$ the ordinary ridge regression estimator of Hoerl and Kennard,

$$p'\hat{\beta} = p'(X'X + kI)^{-1}X'Y, \qquad (4.5.6)$$

is obtained.

Let $G = UKU'$ where K is a diagonal matrix. Then

$$p'\hat{\beta} = p'(X'X + UKU')^+X'Y. \qquad (4.5.7)$$

Let $G = cX'X$ where c is a positive constant. Then

$$p'\hat{\beta} = \frac{p'b}{1+c}. \qquad (4.5.8)$$

This is the contraction estimator of Mayer and Willke (1973). It may be rewritten

$$p'\hat{\beta} = \left[1 - \frac{c}{1+c}\right] p'b. \tag{4.5.9}$$

If $c/(1+c)$ is estimated by $(\hat{\sigma}^2 d)/(b'X'Xb)$, where

$$\hat{\sigma}^2 = \frac{1}{r(n-s)}[Y'Y - Y'Xb] \tag{4.5.10}$$

the usual within sum of squares, then

$$p'\hat{\beta} = \left[1 - \frac{\hat{\sigma}^2 d}{b'X'Xb}\right] p'b. \tag{4.5.11}$$

The estimator in (4.5.11) is the well known James-Stein estimator.
Let $G = UC\Lambda U'$ with C a diagonal matrix. Then

$$p'\hat{\beta} = p'(U\Lambda U' + UC\Lambda U')^+ X'Y$$

$$= p'U(I+C)^{-1}\Lambda^{-1}U'X'Y$$

$$= p'U(I+C)^{-1}U'b \tag{4.5.12}$$

$$= p'[U(I+C)^{-1}U' + VV']b$$

$$= p'[U\ \ V]\begin{bmatrix}(I+C) & 0 \\ 0 & I\end{bmatrix}^{-1}\begin{bmatrix}U' \\ V'\end{bmatrix}b.$$

Thus,

$$p'\hat{\beta} = p' \left[[U \ V] \begin{bmatrix} (I+C) & 0 \\ 0 & I \end{bmatrix} \begin{bmatrix} U' \\ V' \end{bmatrix} \right]^{-1} b$$

(4.5.13)

$$= p'[I + UCU']^{-1}b.$$

Estimator (4.5.13) will be called the generalized contraction estimator. For a replicated experimental design it is possible to obtain various James-Stein type estimators. See Gruber (1979) for more details.

It will now be shown how the above ridge type estimators are special cases of mixed estimators.

Suppose $r = 0$ in the augmented linear model (3.3.1) and $R'R = U \Delta U'$ where Δ is a diagonal matrix, e.g. $\Delta^{-1/2}U'R$ is an orthogonal matrix. Then for $\Delta = kI$, $\Delta = k\Lambda$, $\Delta = K$ and $\Delta = K\Lambda$ (4.5.6), (4.5.7), (4.5.8) and (4.5.13) are respectively obtained. This shows the ridge estimators are special cases of the mixed estimator.

Example 4.5.1. Some Numerical Estimates of Different Ridge Type Estimators. Recall that for Example 1.2.2 the least square estimator was

$$b_{LS} = \begin{bmatrix} 264.0925 \\ 0.3842 \\ 0.3718 \\ -0.0042 \end{bmatrix}.$$

When $k = 1$ the ridge estimator was

$$b_R = \begin{bmatrix} 5.1791 \\ 0.1993 \\ 0.5431 \\ -0.0031 \end{bmatrix}.$$

Now when $k = 1$ the contraction estimator is

$$b_C = \begin{bmatrix} 132.6710 \\ 0.1745 \\ 0.1854 \\ 0.0021 \end{bmatrix}.$$

For the diagonal matrix

$$K = \begin{bmatrix} .0625 & 0 & 0 & 0 \\ 0 & 1 & 0 & 0 \\ 0 & 0 & 4 & 0 \\ 0 & 0 & 0 & 9 \end{bmatrix},$$

the generalized ridge estimator is

$$b_{GR} = \begin{bmatrix} 1.44946 \\ 0.19565 \\ 0.54589 \\ 0.00022 \end{bmatrix},$$

and the generalized contraction estimator is

$$b_{GC} = \begin{bmatrix} 26.4093 \\ 0.1729 \\ 0.1893 \\ 0.0010 \end{bmatrix}.$$

Example 4.5.2.   Different Ridge Estimators. For the linear model in Example 3.1.1

$$X'X = \begin{bmatrix} \frac{1}{\sqrt{2}} & 0 \\ \frac{1}{\sqrt{2}} & 0 \\ 0 & 1 \end{bmatrix} \begin{bmatrix} 3 & 0 \\ 0 & 1 \end{bmatrix} \begin{bmatrix} \frac{1}{\sqrt{2}} & \frac{1}{\sqrt{2}} & 0 \\ 0 & 0 & 1 \end{bmatrix}.$$

Thus,

$$
(X'X)^+ = \begin{bmatrix} \frac{1}{\sqrt{2}} & 0 \\ \frac{1}{\sqrt{2}} & 0 \\ 0 & 1 \end{bmatrix} \begin{bmatrix} \frac{1}{3} & 0 \\ 0 & 1 \end{bmatrix} \begin{bmatrix} \frac{1}{\sqrt{2}} & \frac{1}{\sqrt{2}} & 0 \\ 0 & 0 & 1 \end{bmatrix}
$$

$$
= \begin{bmatrix} \frac{1}{6} & \frac{1}{6} & 0 \\ \frac{1}{6} & \frac{1}{6} & 0 \\ 0 & 0 & 1 \end{bmatrix}.
$$

Thus, another form of the LS is

$$
\begin{bmatrix} b_0 \\ b_1 \\ b_2 \end{bmatrix} = \begin{bmatrix} \frac{1}{3}\left(Y_1 + \frac{Y_2}{\sqrt{2}}\right) \\ \frac{1}{3}\left(Y_1 + \frac{Y_2}{\sqrt{2}}\right) \\ Y_3 \end{bmatrix}.
$$

The reader can easily verify that $\beta_0 + \beta_1 + (1/\sqrt{2})\beta_2$ has the same estimator for both forms of the LS of the vector of parameters.

To obtain the generalized ridge estimator, observe that, for

$$
K = \begin{bmatrix} k_1 & 0 \\ 0 & k_2 \end{bmatrix},
$$

$$
(X'X + UKU') = \begin{bmatrix} \frac{1}{\sqrt{2}} & 0 \\ \frac{1}{\sqrt{2}} & 0 \\ 0 & 1 \end{bmatrix} \begin{bmatrix} 3 + k_1 & 0 \\ 0 & 1 + k_2 \end{bmatrix} \begin{bmatrix} \frac{1}{\sqrt{2}} & \frac{1}{\sqrt{2}} & 0 \\ 0 & 0 & 1 \end{bmatrix}
$$

and

$$(X'X + UKU')^+ = \begin{bmatrix} \frac{1}{\sqrt{2}} & 0 \\ \frac{1}{\sqrt{2}} & 0 \\ 0 & 1 \end{bmatrix} \begin{bmatrix} \frac{1}{3+k_1} & 0 \\ 0 & \frac{1}{1+k_2} \end{bmatrix} \begin{bmatrix} \frac{1}{\sqrt{2}} & \frac{1}{\sqrt{2}} & 0 \\ 0 & 0 & 1 \end{bmatrix}$$

$$= \begin{bmatrix} \frac{1}{2(3+k_1)} & \frac{1}{2(3+k_1)} & 0 \\ \frac{1}{2(3+k_1)} & \frac{1}{2(3+k_1)} & 0 \\ 0 & 0 & \frac{1}{1+k_2} \end{bmatrix}.$$

Thus,

$$\begin{bmatrix} \hat{\beta}_0 \\ \hat{\beta}_1 \\ \hat{\beta}_2 \end{bmatrix} = \begin{bmatrix} \frac{1}{2(3+k_1)}(Y_1 + \frac{Y_2}{\sqrt{2}}) \\ \frac{1}{2(3+k_1)}(Y_1 + \frac{Y_2}{\sqrt{2}}) \\ \frac{Y_3}{1+k_2} \end{bmatrix}.$$

For the estimable parametric function $\hat{\beta}_0 + \hat{\beta}_1 + \frac{1}{\sqrt{2}}\hat{\beta}_2$,

$$\hat{\beta}_0 + \hat{\beta}_1 + \frac{1}{\sqrt{2}}\hat{\beta}_2 = \frac{1}{(3+k_1)}\left(Y_1 + \frac{Y_2}{\sqrt{2}}\right) + \frac{1}{\sqrt{2}}\frac{Y_3}{1+k_2}.$$

To obtain the ordinary ridge estimator simply let $k_1 = k_2$. Observe that

$$\hat{\beta}_0 + \hat{\beta}_1 + \hat{\beta}_2 = \frac{3(b_0 + b_1)/2}{(3+k_1)} + \frac{b_3}{\sqrt{2}(1+k_1)}.$$

For the generalized contraction estimator

$$(X'X + UK\Lambda U') = \begin{bmatrix} \frac{1}{\sqrt{2}} & 0 \\ \frac{1}{\sqrt{2}} & 0 \\ 0 & 1 \end{bmatrix} \begin{bmatrix} 3+3k_1 & 0 \\ 0 & 1+k_2 \end{bmatrix} \begin{bmatrix} \frac{1}{\sqrt{2}} & \frac{1}{\sqrt{2}} & 0 \\ 0 & 0 & 1 \end{bmatrix}$$

and

$$(X'X + UK\Lambda U')^+ = \begin{bmatrix} \frac{1}{\sigma^2(1+k_1)} & \frac{1}{\sigma^2(1+k_1)} & 0 \\ \frac{1}{\sigma^2(1+k_1)} & \frac{1}{\sigma^2(1+k_1)} & 0 \\ 0 & 0 & \frac{1}{1+k_2} \end{bmatrix}.$$

Thus

$$\begin{bmatrix} \beta_0^{GC} \\ \beta_1^{GC} \\ \beta_2^{GC} \end{bmatrix} = \begin{bmatrix} \frac{1}{3(1+k_1)}(Y_1 + \frac{Y_2}{\sqrt{2}}) \\ \frac{1}{3(1+k_1)}(Y_1 + \frac{Y_2}{\sqrt{2}}) \\ \frac{Y_3}{1+k_2} \end{bmatrix} = \begin{bmatrix} \frac{1}{1+k_1}b_0 \\ \frac{1}{1+k_1}b_1 \\ \frac{1}{1+k_2}b_2 \end{bmatrix}.$$

The contraction estimator is obtained by letting $k_1 = k_2$.

Exercise 4.5.1. For the model

$$Y = \begin{bmatrix} 1 & 0 & 0 & 1 & 0 & 0 \\ 1 & 0 & 0 & 0 & 1 & 0 \\ 1 & 0 & 0 & 0 & 0 & 1 \\ 0 & 1 & 0 & 1 & 0 & 0 \\ 0 & 1 & 0 & 0 & 1 & 0 \\ 0 & 1 & 0 & 0 & 0 & 1 \\ 0 & 0 & 1 & 1 & 0 & 0 \\ 0 & 0 & 1 & 0 & 1 & 0 \\ 0 & 0 & 1 & 0 & 0 & 1 \end{bmatrix} \begin{bmatrix} \alpha_1 \\ \alpha_2 \\ \alpha_3 \\ \beta_1 \\ \beta_2 \\ \beta_3 \end{bmatrix} + \epsilon$$

formulate the LS, ridge, generalized, ridge contraction and generalized contraction estimator. Give an example of at least one estimable parametric function.

4.6 Extension of Results to the Case Where $U'FU$ Is Not Positive Definite

4.6.1 Introduction and Motivation

The results about the BE and the mixed estimator in Sections 4.1-4.5 rested on the assumption that $U'FU$ is positive definite or $R = AU'$ with A non-singular. This means that for the BE, F must have rank at least s and for the mixed estimator there must be at least $t \geq s$ additional observations.

   In this section the case where the rank of F is less than s or when there are only a few additional observations is considered. For the BE a condition on the prior mean is obtained which enables:

1. alternative forms of the BE to be obtained;
2. it to be shown how the BE is a CLS or a generalized ridge estimator.

The relationship to the results of the preceeding section will be explained.

4.6.2 The Equivalence of the Bayes Estimator and the Mixed Estimator When $U'FU$ Is Not Positive Definite

It will be shown that there exists a reparametization of the augmented linear model where the Bayes and the mixed estimator are equivalent. For this purpose the singular value decomposition

$$U'FU = [P_1 \ Q_1] \begin{bmatrix} \Delta & 0 \\ 0 & 0 \end{bmatrix} \begin{bmatrix} P_1' \\ Q_1' \end{bmatrix} = P_1 \Delta P_1' \qquad (4.6.1)$$

is needed. The reparametization is given by the following theorem.

Theorem 4.6.1. *Let $p'\hat{\beta}$ be the BE of a parametric function $p'\beta$.*

Assume:

1. The parametric function $p'\beta$ is estimable.
2. $p'UP_1 \neq 0$.
3. The vector $U'\theta$ is in the column space of P.

Given:

1. the linear model

$$Y = Z\alpha + \epsilon \tag{4.6.2}$$

where $Z = XUP_1$, $\alpha = P_1'U'\beta$, $E(\epsilon|\alpha) = 0$, $D(\epsilon|\alpha) = \sigma^2 I$, and
2. the stochastic prior constraints

$$r = R\alpha + \epsilon \quad \text{where} \quad R'R = \Delta^{-1} \tag{4.6.3a}$$

with $E(\epsilon|\alpha) = 0$ and $D(\epsilon|\alpha) = \tau^2 I$. Let $c'\hat{\alpha}$ be the mixed estimator for the augmented model in (4.6.2) and (4.6.3) where $c = p'U$. Then,

$$p'\hat{\beta} = c'\hat{\alpha}. \tag{4.6.3b}$$

Proof. Since $[P_1 \ Q_1]$ is an orthogonal matrix

$$U'\theta = (P_1 P_1' + Q_1 Q_1')U'\theta = P_1\eta + Q_1\psi, \tag{4.6.4}$$

where

$$\eta = P_1'U'\theta \quad \text{and} \quad \psi = Q_1'U'\theta. \tag{4.6.5}$$

The linear model may be written

$$Y = X\beta + \epsilon = XUU'\beta + \epsilon = XU(P_1 P_1' + Q_1 Q_1')U'\beta + \epsilon$$

$$= XUP_1\alpha + XUQ_1\delta + \epsilon, \tag{4.6.6}$$

where

$$\alpha = P_1'U'\beta \quad \text{and} \quad \delta = Q_1'U'\beta. \tag{4.6.7}$$

The Bayes estimator (3.5.7) may be written

$$p'\hat{\beta} = p'UU'\theta + p'UU'FUU'X'[XUU'FUU'X' + \sigma^2 I]^{-1}(Y - X\theta)$$

$$= [p'UP_1\eta + p'UP_1\Delta P_1'U'X'A^{-1}(Y - XUP_1\eta)]$$

$$+ [p'UQ_1\psi + p'UP_1\Delta P_1'X'A^{-1}XUQ_1\psi],$$

$$\tag{4.6.8}$$

where

$$A = XUP_1\Delta P_1'U'X' + \sigma^2 I.$$

The term in the second set of brackets (4.6.8) is zero if $Q_1\psi = 0$ or equivalently if $U'\theta$ is in the column space of P_1. Then

$$p'\hat{\beta} = c'\eta + c'\Delta Z'[Z\Delta Z' + \sigma^2 I]^{-1}(Y - Z\eta). \tag{4.6.9}$$

Given the prior assumptions

$$E(\alpha) = \eta \quad \text{and} \quad D(\alpha) = \Delta, \tag{4.6.10}$$

Equation (4.6.9) is the Bayes estimator for the model (4.6.2). Since

$$Z'Z = P_1'U'X'XUP_1 = P_1'\Lambda P_1, \tag{4.6.11}$$

$Z'Z$ is positive definite. From the results of Section 3.1,

$$p'\hat{\beta} = c'\hat{\alpha} = c'[Z'Z + \sigma^2\Delta^{-1}]^{-1}[Z'Y + \sigma^2\Delta^{-1}\eta]. \tag{4.6.12}$$

Let R be such that $R'R = \Delta$. Equation (4.6.12) is the mixed estimator sought. ∎

A simple corollary of Theorem 4.6.1 above is that for the models (4.6.2) and (4.6.3) the sample and the prior information are exchangeable.

When $U'FU$ is positive definite, $Q = 0$. The second term in (4.6.8) is zero. Thus, equation (4.6.8) may be written

$$p'\hat{\beta} = p'U P_1 \eta + p'U[\sigma^2 F + U'X'XU]^{-1}U'X'(Y - XU P_1 \eta)$$

$$= p'\theta + p'[\sigma^2(UU'FUU')^+ + X'X]^+ X'(Y - X\theta) \quad (4.6.13)$$

as was derived previously.

The least square estimator of α in (4.6.2) is

$$a = (Z'Z)^{-1} Z'Y.$$

Estimator (4.6.9) is the BE of $c'\alpha$. Thus, from Section 4.1 since $Z'Z$ and Δ are non-singular the alternative forms are

$$c'\hat{\alpha} = c'\eta + c'[Z'Z + \sigma^2 \Delta^{-1}]^{-1} Z'(Y - Z\eta) \quad (4.6.14)$$

$$= c'[Z'Z + \sigma^2 \Delta^{-1}]^{-1}[Z'Y + \Delta^{-1}\sigma^2] \quad (4.6.15)$$

$$= c'\eta + c'\Delta[\sigma^2(Z'Z)^{-1} + \Delta]^{-1}(a - \eta) \quad (4.6.16)$$

$$= c'(Z'Z)^{-1}[\Delta + \sigma^2(Z'Z)^{-1}]^{-1}\eta\sigma^2$$

$$+ c'\Delta[\Delta + \sigma^2(Z'Z)^{-1}]^{-1}a \quad (4.6.17)$$

$$= c'a - c'(Z'Z)^{-1}[\Delta + \sigma^2(Z'Z)^{-1}]^{-1}(a - \eta)\sigma^2. \quad (4.6.18)$$

The interpretations of these forms are similar to those in Sections 4.1 and 4.2.

Example 4.6.1. Illustrations of Theorem 4.6.1. Consider Model 1 in Example 4.4.1. Assume that the prior mean and dispersion are

$$\theta = \begin{bmatrix} 1 \\ 1 \\ 1 \\ 1 \end{bmatrix} \quad \text{and} \quad F = \begin{bmatrix} 1 & 0 & 0 & 0 \\ 0 & 1 & 0 & 0 \\ 0 & 0 & \frac{1}{2} & \frac{1}{2} \\ 0 & 0 & \frac{1}{2} & \frac{1}{2} \end{bmatrix}.$$

The problem is to find the full rank reparametization of the model and the parametric function $c'\alpha$ where the BE of $\mu + \tau_3$, $c'\hat{\alpha}$ is the corresponding mixed estimator. First obtain $U'FU$ and its SVD. Thus,

$$U'FU = \begin{bmatrix} 1 & 0 & 0 \\ 0 & 0 & 0 \\ 0 & 0 & 1 \end{bmatrix} = \begin{bmatrix} 1 & 0 & 0 \\ 0 & 0 & 1 \\ 0 & 1 & 0 \end{bmatrix} \begin{bmatrix} 1 & 0 & 0 \\ 0 & 1 & 0 \\ 0 & 0 & 0 \end{bmatrix} \begin{bmatrix} 1 & 0 & 0 \\ 0 & 0 & 1 \\ 0 & 1 & 0 \end{bmatrix}.$$

Hence

$$P_1 = \begin{bmatrix} 1 & 0 \\ 0 & 0 \\ 0 & 1 \end{bmatrix}, \quad Q_1 = \begin{bmatrix} 0 \\ 1 \\ 0 \end{bmatrix}, \quad \text{and} \quad \Delta = \begin{bmatrix} 1 & 0 \\ 0 & 1 \end{bmatrix}.$$

Also

$$Z = XUP' = \begin{bmatrix} \frac{4}{\sqrt{12}} \cdot 1_3 & \frac{-2}{\sqrt{6}} \cdot 1_3 \\ \frac{4}{\sqrt{12}} \cdot 1_3 & \frac{1}{\sqrt{6}} \cdot 1_3 \\ \frac{4}{\sqrt{12}} \cdot 1_3 & \frac{1}{\sqrt{6}} \cdot 1_3 \end{bmatrix}.$$

Thus, the augmented model (4.6.2) is

$$
Y = \begin{bmatrix}
\frac{4}{\sqrt{12}} \cdot 1_3 & -\frac{2}{\sqrt{6}} \cdot 1_3 \\[6pt]
\frac{4}{\sqrt{12}} \cdot 1_3 & \frac{1}{\sqrt{6}} \cdot 1_3 \\[6pt]
\frac{4}{\sqrt{12}} \cdot 1_3 & \frac{1}{\sqrt{6}} \cdot 1_3 \\[6pt]
\frac{1}{\sqrt{2}} & -\frac{1}{\sqrt{2}} \\[6pt]
\frac{1}{\sqrt{2}} & \frac{1}{\sqrt{2}}
\end{bmatrix}
\begin{bmatrix} \alpha_1 \\ \alpha_2 \end{bmatrix} + \epsilon.
$$

For R any orthogonal matrix could be used. Now

$$
c' = p'U P_1
$$

from Equation (4.6.8). Thus

$$
c' = [1\ 0\ 0\ 1]
\begin{bmatrix}
\frac{3}{\sqrt{12}} & 0 & 0 \\[6pt]
\frac{1}{\sqrt{12}} & 0 & -\frac{2}{\sqrt{6}} \\[6pt]
\frac{1}{\sqrt{12}} & -\frac{1}{\sqrt{2}} & \frac{1}{\sqrt{6}} \\[6pt]
\frac{1}{\sqrt{12}} & \frac{1}{\sqrt{2}} & \frac{1}{\sqrt{6}}
\end{bmatrix}
\begin{bmatrix} 1 & 0 \\ 0 & 0 \\ 0 & 1 \end{bmatrix}
= [\ \frac{4}{\sqrt{12}}\ \ \frac{1}{\sqrt{6}}\ ].
$$

Now $\widehat{u + \tau_3} = \dfrac{4}{\sqrt{12}} \hat{\alpha}_1 + \dfrac{1}{\sqrt{6}} \hat{\alpha}_2.$

Exercise 4.6.1. For Example 4.6.1
   A. Find the least square estimator of α.
   B. Find the parametric function $c'\alpha$ where for the BE of $\tau_1 - \tau_3$, $c'\alpha$ is the corresponding mixed estimator.
   C. Show that $\theta = 1_4$ is in the column space of P.

D. Show that for $\tau_2 - \tau_3$, $p'UP_1 = 0$, i.e. Condition 2 of Theorem 4.6.1 is violated.

Exercise 4.6.2. A.Where in the proof of Theorem 4.6.1 is estimability used? B.Is estimability really needed? Can the theorem be proved without assumption 1 or with something else in its place?

Exercise 4.6.3. In Theorem 4.6.1 let

$$X = \begin{bmatrix} 2 & -2 & 0 & 0 & 0 \\ 1 & 1 & -2 & 0 & 0 \\ \frac{1}{2} & \frac{1}{2} & \frac{1}{2} & -\frac{3}{2} & 0 \\ \frac{1}{10} & \frac{1}{10} & \frac{1}{10} & \frac{1}{10} & -\frac{4}{10} \end{bmatrix},$$

$$R = \begin{bmatrix} \frac{1}{\sqrt{2}} & \frac{1}{\sqrt{2}} & 0 \\ \frac{1}{\sqrt{2}} & -\frac{1}{\sqrt{2}} & 0 \\ 0 & 0 & \sqrt{\frac{5}{2}} \end{bmatrix}.$$

A. Interpret $U'\theta$ is in column space of P.
B. Find Z and the transformation equations that link $(\alpha_1, \alpha_2, \alpha_3)$ with $(\beta_1, \beta_2, \beta_3, \beta_4, \beta_5)$.
C. What is the prior mean η in terms of θ where $\theta' = (\theta_1, \theta_2, \theta_3, \theta_4, \theta_5)$?
D. Is $\beta_1 - \beta_2$ is estimable? If it is, what parametric function $c'\alpha$ is such that $\beta_1 - \beta_2 = c'\alpha$?

The discussion below requires the SVD of

$$R = [A' \ B'] \begin{bmatrix} \Theta^{1/2} & 0 \\ 0 & 0 \end{bmatrix} \begin{bmatrix} P_2' \\ Q_2' \end{bmatrix}. \qquad (4.6.19)$$

Theorem 4.6.1 starts out with the Bayes estimator. It gives a reparametization of the linear model where the Bayes estimator is equivalent to a mixed estimator. Theorem 4.6.2 below starts out with the mixed estimator and explains how to reparametize the linear model so that it is a special case ot the Bayes estimator.

Theorem 4.6.2. *Assume matrix R in $r = R\beta + \theta$ had rank less than s. Restrict the β parmameters to the subspace where $Q_2'\beta = 0$. The model*

$$\begin{bmatrix} Y \\ r \end{bmatrix} = \begin{bmatrix} X \\ R \end{bmatrix} \beta + \begin{bmatrix} \epsilon \\ \phi \end{bmatrix} \tag{4.6.20}$$

can be reparametized and a prior distribution found where the mixed estimator $p'\hat{\beta}$ is Bayes.

Proof. When $Q_2'\beta = 0$, (4.6.20) may be written

$$\begin{bmatrix} Y \\ r \end{bmatrix} = \begin{bmatrix} X P_2 P_2' \\ R P_2 P_2' \end{bmatrix} \beta + \begin{bmatrix} \epsilon \\ \phi \end{bmatrix} \tag{4.6.21}$$

since $RQ_2 = 0$. Let $\gamma = P_2'\beta$, $M = XP_2$ and $N = RP_2$. Rewrite (4.6.20) as

$$\begin{bmatrix} Y \\ r \end{bmatrix} = \begin{bmatrix} M \\ N \end{bmatrix} \gamma + \begin{bmatrix} \epsilon \\ \phi \end{bmatrix}. \tag{4.6.22}$$

Now $N'N = P_2'R'RP_2 = \Delta$. Then

$$p'\hat{\gamma} = p' \left(M'M\tau^2 + \Delta\sigma^2 \right)^{-1} \left(M'Y\tau^2 + N'r\sigma^2 \right). \tag{4.6.23}$$

Since Δ is positive definite, (4.6.23) is an alternative form of the Bayes estimator for the prior with mean and dispersion

$$E(\gamma) = \hat{\gamma}_2 = \Delta^{-1} P_2' N'r \quad \text{and} \quad D(\gamma) = \Delta^{-1}\sigma^2. \tag{4.6.24}$$

The reparametized linear model is

$$Y = M\gamma + \epsilon. \ \blacksquare \tag{4.6.25}$$

Remark. A similar representation may be found for the case where $V'\beta = 0$.

In Theorem 4.6.2 the hypothesis $Q_2'\beta = 0$ could have been replaced by $XQ_2 = 0$. However $XQ_2 = 0$ implies that $U'Q_2 = 0$. Thus,

$$Q_2 = (UU' + VV')Q_2 = VV'Q_2. \tag{4.6.26}$$

Now

$$m - t = \text{rank } Q_2 \leq \text{rank } VV' = m - s. \tag{4.6.27}$$

Thus $t \geq s$. Then R is of rank at least s given that $XQ = 0$.

Example 4.6.2.   Illustration of Theorem 4.6.2. Consider an augmented model of the form (4.6.20) where X is as in Example 4.6.1 and

$$R = \begin{bmatrix} 1 & 1 & 1 & 1 \\ 0 & 0 & 1 & -1 \end{bmatrix}.$$

The reparametization of the model will be exhibited and the prior that $p'\hat{\beta}$ is Bayes with respect to obtained. The technique is to obtain first the SVD of $R'R$, then M, N, $\hat{\gamma}$ and Δ^{-1} as defined in the proof of Theorem 4.6.2. First observe

$$R'R = \begin{bmatrix} 1 & 1 & 1 & 1 \\ 1 & 1 & 1 & 1 \\ 1 & 1 & 2 & 0 \\ 1 & 1 & 0 & 2 \end{bmatrix}$$

$$= [P_2 \ Q_2] \begin{bmatrix} 4 & 0 & 0 & 0 \\ 0 & 2 & 0 & 0 \\ 0 & 0 & 0 & 0 \\ 0 & 0 & 0 & 0 \end{bmatrix} \begin{bmatrix} P_2' \\ Q_2' \end{bmatrix},$$

with

$$P_2 = \begin{bmatrix} \frac{1}{2} & 0 \\ \frac{1}{2} & 0 \\ \frac{1}{2} & \frac{1}{\sqrt{2}} \\ \frac{1}{2} & -\frac{1}{\sqrt{2}} \end{bmatrix}, \qquad Q_2 = \begin{bmatrix} 0 & -\frac{3}{\sqrt{12}} \\ -\frac{2}{\sqrt{6}} & \frac{1}{\sqrt{12}} \\ \frac{1}{\sqrt{6}} & \frac{1}{\sqrt{12}} \\ \frac{1}{\sqrt{6}} & \frac{1}{\sqrt{12}} \end{bmatrix}.$$

Now

$$M = XP_2 = \begin{bmatrix} 1_3 & 0 \\ 1_3 & \frac{1}{\sqrt{2}} \cdot 1_3 \\ 1_3 & -\frac{1}{\sqrt{2}} \cdot 1_3 \end{bmatrix}$$

and

$$N = RP_2 = \begin{bmatrix} 2 & 0 \\ 0 & \sqrt{2} \end{bmatrix}.$$

Thus,

$$N'N = \begin{bmatrix} 4 & 0 \\ 0 & 2 \end{bmatrix} = \Delta.$$

The subspace where $Q_2'\beta = 0$ is defined by

$$-2\tau_1 + \tau_2 + \tau_3 = 0,$$
$$-3\mu + \tau_1 + \tau_2 + \tau_3 = 0.$$

From Equation (4.6.24) the prior mean is

$$\hat{\gamma}_2 = \begin{bmatrix} \frac{1}{4} & 0 \\ 0 & \frac{1}{2} \end{bmatrix} \begin{bmatrix} \frac{1}{2} & \frac{1}{2} & \frac{1}{2} & \frac{1}{2} \\ 0 & 0 & \frac{1}{\sqrt{2}} & -\frac{1}{\sqrt{2}} \end{bmatrix} \begin{bmatrix} 1 & 0 \\ 1 & 0 \\ 1 & 1 \\ 1 & -1 \end{bmatrix} \begin{bmatrix} r_1 \\ r_2 \end{bmatrix} = \begin{bmatrix} \frac{1}{2}r_1 \\ \frac{1}{\sqrt{2}}r_2 \end{bmatrix}$$

and the prior dispersion is

$$D(\gamma) = \begin{bmatrix} \frac{1}{2} & 0 \\ 0 & \frac{1}{2} \end{bmatrix} \sigma^2.$$

From Equation (4.6.25) the reparametrized model is

$$Y = \begin{bmatrix} 1_3 & 0 \\ 1_3 & \frac{1}{\sqrt{2}} \cdot 1_3 \\ 1_3 & -\frac{1}{\sqrt{2}} \cdot 1_3 \end{bmatrix} \gamma + \epsilon.$$

Exercise 4.6.4. In Theorem 4.6.2 let

$$X = \begin{bmatrix} 1_4 & 1_4 & 0 & 0 \\ 1_4 & 0 & 1_4 & 0 \\ 1_4 & 0 & 0 & 1_4 \end{bmatrix}$$

and

$$R = R_2 = \begin{bmatrix} 0 & 1 & -1 & 0 \\ 0 & 1 & 1 & -2 \end{bmatrix}$$

A. Characterize the subspace where $Q_2'\beta = 0$.
B. Find M, N, Δ.
C. What is γ in terms of β?
D. What are the prior mean and variance in (4.6.24) and the resulting reparameterized full rank linear model?

Exercise 4.6.5. For the X matrix in Exercise 4.6.4 above redo the problem when

$$R = R_1 = \begin{bmatrix} 0 & 1 & -1 & 0 \end{bmatrix},$$

$$R = R_3 = \begin{bmatrix} 0 & 1 & -1 & 0 \\ 0 & 1 & 1 & -2 \\ 1 & 1 & 1 & 1 \end{bmatrix}.$$

4.6.3 The Generalized Ridge Regression Estimator

The mixed estimator in Equation (4.6.12) may also be obtained as a form of the generalized ridge regression estimator. The technique is the same as that used in Section 3.2, but the parameters and the design matrix are those of Section 4.6.1.

The estimator is obtained by minimizing the expression

$$m = (\alpha - \eta)'\Delta^{-1}(\alpha - \eta) \tag{4.6.28}$$

subject to the side condition

$$(\alpha - a)'Z'Z(\alpha - a) = \Phi_0, \tag{4.6.29}$$

where $a = (Z'Z)^{-1}Z'Y$ (the least square estimator for the model (4.6.2)). As was the case in Section 3.2 and Hoerl and Kennard (1970) the method of Lagrange multipliers is employed. Differentiate

$$L = (\alpha - \eta)'\Delta^{-1}(\alpha - \eta) + \frac{1}{k}(\alpha - a)'Z'Z(\alpha - a) \tag{4.6.30}$$

to obtain

$$\Delta^{-1}(\alpha - \eta) + \frac{1}{k}Z'Z(\alpha - a) = 0. \qquad (4.6.31)$$

Thus,

$$\hat{\alpha} = (\Delta^{-1} + \frac{1}{k}Z'Z)^{-1}(\frac{1}{k}Z'Za + \Delta^{-1}\eta)$$

$$\qquad (4.6.32)$$

$$= (k\Delta^{-1} + Z'Z)^{-1}(Z'Za + k\Delta^{-1}\eta).$$

The form of (4.6.32) and (4.6.15) are the same with k replacing σ^2. Consequently, with respect to the original model

$$Y = X\beta + \epsilon$$

and the prior assumptions

$$E(\beta) = \theta \quad \text{and} \quad D(\beta) = \frac{1}{k}F, \qquad (4.6.33)$$

where $F = P\Delta^{-1}P'$, the estimator (4.6.32) is Bayes.

4.7 An Extension of the Gauss-Markov Theorem

In Section 3.1 it was observed that the least square estimator was a minimum variance unbiased estimator. This was the content of the Gauss-Markov Theorem.

In Section 3.5 for the prior assumptions (3.5.1) the linear estimator with minimum variance subject to (3.5.4) was obtained. Equation (3.5.4) may be rewritten

$$E(a + L'Y) = p'\theta. \qquad (4.7.1)$$

Thus, $a + L'Y$ can be thought of as an unbiased linear estimator of θ. Assuming that β is constant is equivalent to assuming that β is a random variable with degenerate distribution i.e. a distribution with zero variance.

The statements of the preceding paragraph above motivates the following definitions:

Definition 4.7.1. A linear estimator $a + L'Y$ is an extended unbiased estimator of $p'\beta$ if its the expectation averaged over the prior distribution is equal to the prior mean.

Definition 4.7.2. A linear estimator $a + L'Y$ is MVUE (a minimum variance unbiased estimator) of $p'\beta$ if it is a solution to the optimization problem in (3.5.3) and (3.5.4).

In Sections 4.4 and 4.6 conditions were given for the Bayes estimator and the mixed estimator to be equivalent using the alternative forms of the Bayes estimator. These results may be summarized in the extended Gauss-Markov theorem stated below.

Theorem 4.7.1.
1. *The linear BE is MVUE.*
2. *When $U'FU$ is positive definite there is an augmented linear model*

$$\begin{bmatrix} Y \\ r \end{bmatrix} = \begin{bmatrix} X \\ R \end{bmatrix} \beta + \begin{bmatrix} \epsilon \\ \eta \end{bmatrix} \tag{4.7.2}$$

such that the linear BE is the LS estimator.
3. *When $U'FU$ has rank less that of equal to s (this includes both the case where $U'FU$ is positive definite and when it is not) there is a reparameterized augmented linear model of full rank*

$$\begin{bmatrix} Y \\ r \end{bmatrix} = \begin{bmatrix} Z \\ R \end{bmatrix} \alpha + \begin{bmatrix} \epsilon \\ \eta \end{bmatrix} \tag{4.7.3}$$

such that $p'\hat{\beta} = c'\hat{\alpha}$ where $p'\hat{\beta}$ is the linear BE and $p'\beta$ and $c'\hat{\alpha}$ is the LS estimator with respect to (4.7.3).

4.8 Summary and Remarks

In this chapter

1. Alternative forms of the Bayes estimator were derived. The cases considered included:
   a. X and F of full rank;
   b. X of non-full rank $U'FU$ positive definite;
   c. X of non-full rank $U'FU$ positive semidefinite.
2. Conditions were obtained for the generalized ridge estimator and the Bayes estimator to be equivalent.
3. Conditions were obtained for the mixed estimator and the Bayes estimator to be equivalent. The cases considered were
   a. $R = AU'$ with A a PD matrix;
   b. $R'R$ was positive semidefinite.
4. Situations whose sample and prior information could be exchanged were investigated.
5. How the ridge estimators in the literature were special cases of the Bayes estimator was taken up.

It was observed that the form of the Bayes estimator for the non-full rank model in general was the same as that of the full rank case with Moore-Penrose inverses replacing ordinary inverses. Results on the equivalence of BE, mixed and ridge estimators, and exchangeability of sample and prior information generally held true when $U'FU$ was positive semidefinite. When $U'FU$ was positive semidefinite of non-full rank equivalences among estimators could be established if one of the following two things were done.

1. The β or θ parameters were restricted to the appropriate subspace.
2. An appropriate reparametization of the linear model to one of full rank was found.

PART III

THE EFFICIENCIES
OF THE ESTIMATORS

Chapter V

Measures of Efficiency of the Estimators

The criteria for comparison of the estimators to be used in the rest of the book will now be presented. These include the variance, average and conditional MSE and the MSE averaging over a quadratic loss function.

Three kinds of prior information about β were considered in Chapters III and IV.

1. The parameter β was a random variable with

$$E(\beta) = \theta \quad \text{and} \quad D(\beta) = F. \tag{5.1}$$

2. The linear model was augmented by the model

$$r = R\beta + \varepsilon \tag{5.2}$$

with

$$E(\varepsilon) = 0 \text{ and } D(\varepsilon) = \tau^2 I. \tag{5.3}$$

The parameter β was a constant. The same estimators resulted however for parametric functions that were estimable with respect to the augmented model with β a random variable satisfying (5.1) with θ unknown.

151

3. The parameter β was assumed to be a constant lying on the ellipsoid

$$(\beta - \theta)' F^+ (\beta - \theta) \leq 1. \tag{5.4}$$

Three forms of the MSE for the parametric functions $p'\beta$ will be considered in the remainder of this book. These include:

1. the MSE averaging over the prior assumptions (5.1), (average MSE);
2. the MSE without averaging over prior assumptions (conditional MSE);
3. the MSE when the prior assumptions (5.1) are incorrect.

The average MSE is denoted by

$$m_1 = p' E(\hat{\beta} - \beta)(\hat{\beta} - \beta)' p. \tag{5.5}$$

The conditional MSE is denoted by

$$m_2 = p' E_\beta (\hat{\beta} - \beta)(\hat{\beta} - \beta)' p. \tag{5.6}$$

When prior information of the form of (5.1) is available $m_2 = E(m_1)$. If β is a constant the subscripted β in Equation (5.6) is not needed. Suppose that in place of (5.1) the "correct" prior assumptions are

$$E(\beta) = \eta \quad \text{and} \quad D(\beta) = H. \tag{5.7}$$

The MSE averaging over the "correct" prior assumptions is denoted by

$$m_3 = p' E_{\eta,H} (\hat{\beta} - \beta)(\hat{\beta} - \beta)' p. \tag{5.8}$$

Likewise if the ellipsoid (5.4) is incorrect and the correct ellipsoid is

$$(\beta - \eta)' H^+ (\beta - \eta) \leq 1 \tag{5.9}$$

the MSE is denoted by

$$m_4 = p' E_{\beta,\eta,H} (\hat{\beta} - \beta)(\hat{\beta} - \beta)' p. \tag{5.10}$$

The MSE m_2 is the sum of the conditional variance

$$v = p' E_\beta[\hat{\beta} - E_\beta(\hat{\beta})][\hat{\beta} - E_\beta(\hat{\beta})]'p \tag{5.11}$$

and the squared bias

$$a = p'[\beta - E_\beta(\hat{\beta})][\beta - E_\beta(\hat{\beta})]'p. \tag{5.12}$$

Thus,

$$m_2 = E(v) + E(a). \tag{5.13}$$

The MSE in (5.5), (5.6) and (5.8) can also be obtained averaging over a quadratic loss function. Let M be a non-negative definite matrix. The MSE averaging over a quadratic loss function and the prior assumptions (5.1) is

$$M_1 = E(\hat{\beta} - \beta)'M(\hat{\beta} - \beta). \tag{5.14}$$

The MSE without averaging over the prior assumptions is

$$M_2 = E_\beta(\hat{\beta} - \beta)'M(\hat{\beta} - \beta). \tag{5.15}$$

When the prior assumptions in (5.1) are incorrect and the correct prior assumptions are given by (5.7),

$$M_{3,\eta,H} = E_{\eta,H}(\hat{\beta} - \beta)'M(\hat{\beta} - \beta). \tag{5.16}$$

From Theobald's result Theorem 2.4.1 $p\hat{\beta}_1$ has a smaller MSE than $p'\hat{\beta}_2$ for each MSE (5.5), (5.6) and (5.8) if and only if the corresponding MSE in (5.14)–(5.16) is smaller.

The reader will recall from Section 2.2 that the average MSE is the Bayes risk. Also when the loss function is quadratic the average MSE of the Bayes estimator is the variance of the posterior distribution.

The computation, properties, and the comparison of m_1, m_2, m_3 and m_4 will each be taken up in Chapters VI, VII and VIII.

Example 5.1. MSE of Ordinary Ridge Estimator. For the ordinary ridge estimator

$$p'\hat{\beta} = p'(X'X + kI)^{-1}X'Y$$

the variance is

$$v = p'(X'X + kI)^{-1}X' \text{ Var}(YY')X(X'X + kI)^{-1}p$$

$$= p'(X'X + kI)^{-1}X'X(X'X + kI)^{-1}p\sigma^2.$$

Since

$$E(p'\beta) = p'(X'X + kI)^{-1}X'X\beta,$$

the squared bias

$$a = p'[I - (X'X + kI)^{-1}X'X]\beta\beta'[I - X'X(X'X + kI)^{-1}]p$$

$$= p'(X'X + kI)^{-1}\beta\beta'(X'X + kI)^{-1}pk^2.$$

Thus, employing (5.1.3),

$$m_2 = p'(X'X + kI)^{-1}[\sigma^2 X'X + k^2\beta\beta'](X'X + kI)^{-1}p.$$

Recall that the ridge estimator was derived as a BE for

$$E(\beta) = 0 \quad \text{and} \quad D(\beta) = \frac{\sigma^2}{k}I.$$

Now

$$m_1 = E(m_2) = p'(X'X + kI)^{-1}p.$$

Example 5.2. MSE of Contraction Estimator Averaging Over a Quadratic Loss Function. Assume X is of full rank.

In (5.14) let $M = X'X$. Then for the contraction estimator

$$p'\hat{\beta} = \frac{1}{1+k}p'b = (1 - \frac{k}{1+k})p'b,$$

$$E(\hat{\beta} - \beta)'(X'X)(\hat{\beta} - \beta)$$
$$= E(b - \beta)'X'X(b - \beta) - \frac{2k}{1+k}E(b'X'X(b - \beta))$$
$$+ \frac{k^2}{(1+k)^2}E(b'X'Xb)$$
$$= \sigma^2 m - \frac{2k}{1+k}[\sigma^2 m] + \frac{k^2}{(1+k)^2}[\sigma^2 m + \beta'X'X\beta]$$
$$= [1 - \frac{k}{1+k}]^2\sigma^2 m + \frac{k^2}{(1+k)^2}\beta'X'X\beta$$
$$= \frac{1}{(1+k)^2}[\sigma^2 m + k^2\beta'X'X\beta].$$

Example 5.3. MSE of Ordinary Ridge Estimator Averaging Over a Non-Zero Prior Mean When a Zero Prior Mean Is Incorrect. Suppose that in Example 5.1 above

$$E(\beta) = \theta \quad \text{and} \quad D(\beta) = \frac{\sigma^2}{k}I$$

are the correct prior assumptions. Then

$$m_4 = p'(X'X + kI)^{-1}p\sigma^2 + p'(X'X + kI)^{-1}\theta\theta'(X'X + kI)^{-1}pk^2.$$

Exercises - Chapter V

Exercise 5.1. A. For the BE in Exercise 3.5.2 find the MSE

(1) Averaging over the prior.

(2) Without averaging over the prior.

B. Show that the average MSE is less than that of the LS estimator and find the range of values of μ where the conditional MSE is less than that of the LS estimator.

Exercise 5.2. Let $y_{ij} = \mu_i + \varepsilon_{ij}$, $1 \leq i \leq r$, $1 \leq j \leq n$ be r independent linear models with prior assumptions as in Exercise 3.5.2. Suppose θ, σ^2 and τ^2 are unknown.

A. Show that

$$\hat{\mu}_* = \frac{1}{nr} \sum_{i=1}^{r} \sum_{j=1}^{n} y_{ij}$$

$$B = \frac{1}{r} \sum_{i=1}^{r} (\hat{\mu}_i - \hat{\mu}_*)^2,$$

where $\hat{\mu}_i$ is the LS of μ_i and $\hat{\sigma}^2 = 1/(n-r) \sum_{i=1}^{r} (Y_i'Y_i - Y_i'X\hat{\mu}_i)$, are unbiased estimators of $\theta, (\sigma^2/n) + \tau^2$ and σ^2 respectively.

B. Formulate the empirical BE by substituting the estimates above for the parameters of the BE.

C. Obtain the MSE of the estimator for optimum c with and without averaging over the prior.

$$\text{MSE} = E \sum_{i=1}^{r} (\mu_i^{(eb)} - \mu_i)^2.$$

[Hint: Recall formulation of JS estimator to do B.]

Exercise 5.3. Find the MSE of the generalized ridge and generalized contraction estimators in Section 4.5 with and without averaging over the prior distribution. Also find the variance and bias terms.

Exercise 5.4. Find the MSE of the James-Stein estimator when $M = X'X$ for the optimal value of c. What do you notice about how this estimator's MSE compares with that of the LS estimator?

Exercise 5.5. Suppose that in Exercise 3.5.2 the prior assumptions are incorrect. The correct prior assumptions are

$$E(\mu) = \eta, \quad D(\mu) = \delta^2.$$

A. Find the MSE of the BE obtained in Exercise 3.5.2 for the correct prior assumptions.

B. How does this MSE compare with that of the LS estimator?

Chapter VI

The Average MSE

6.0 Introduction

The main purpose of this chapter is to obtain explicit forms of the measure of efficiency of the minimax, the Bayes and the mixed estimators and to show how these forms depend on the prior information. The measure of efficiency of the linear minimax estimator is the minimum value of the maximum risk of the linear estimators on an ellipsoid. The efficiency of the Bayes estimator is measured by its mean square error averaging over a prior distribution. The mixed estimator's efficiency is measured by its variance. The mathematical form of these measures is given in Section 6.1.

In Section 6.2 it is shown that the average MSE of the BE is the difference between the variance of the prior distribution and the average variance of the BE. This fact is used in Section 6.3 to obtain equivalent forms of the MSE for the different alternative forms of the BE. These forms illustrate how the MSE of the BE is smaller than that of the least square estimator and the variance of the prior distribution.

The algebraic forms resulting from calculation of the measures of efficiency of the Bayes and the minimax estimators are the same. The corresponding alternative forms of the variance of the mixed estimator are found in Section 6.4. These forms show how the mixed estimator has a smaller variance than the least square estimator for both model I and model II.

Intuitively, one would expect that more precise knowledge of the prior distribution would produce a BE with a smaller MSE. This is indeed the case as is shown in Theorem 6.5.1 in Section 6.5. The result of this theorem is used in Section 6.6 to give conditions on the prior parameters for one ridge type estimator to have a smaller MSE than another one.

6.1 The Forms of the MSE for the Minimax, Bayes and the Mixed Estimator

In Section 3.4 the optimum estimator (minimax) was found by minimizing the risk

$$p'Rp = (L'X - p')(\beta - \theta)(\beta - \theta)'(X'L - p) + \sigma^2 L'L \qquad (6.1.1)$$

subject to (5.4). In Section 3.5 the optimum estimator (Bayes) was found by minimizing the expectation of (6.1.1) averaging over the prior assumptions (5.1). The two optimum estimators have the same measure of efficiency. Thus, when β lies on the ellipsoid

$$(\beta - \theta)'F^+(\beta - \theta) \le 1,$$

$$m_1 = \max m_2 = p'E(\hat{\beta} - \beta)(\hat{\beta} - \beta)p$$

$$= p'[I - FX'(XFX' + \sigma^2 I)^{-1}X]F[I - X'(XFX' + \sigma^2 I)^{-1}XF]p$$
$$\qquad\qquad\qquad\qquad\qquad\qquad\qquad\qquad\qquad (6.1.2)$$
$$+ p'FX'(XFX' + \sigma^2)^{-2}XFp\sigma^2.$$

Since the LS estimator is unbiased for estimable parametric functions its conditional and average MSE is the same as its conditional variance. Thus,

$$m_1 = m_2 = V(p'\beta) = V(p'(X'X)^+X'Y)$$

$$= p'(X'X)^+X'V(Y)X(X'X)^+p$$

$$= \sigma^2 p'(X'X)^+X'X(X'X)^+p \tag{6.1.3}$$

$$= p'(X'X)^+p\sigma^2.$$

For nonestimable functions the conditional MSE is the sum of the conditional variance and bias. Thus,

$$m_2 = p'(X'X)^+p\sigma^2 + p'VV'(\beta - \theta)(\beta - \theta)'VV'p \tag{6.1.4}$$

and

$$m_1 = p'(X'X)^+p\sigma^2 + p'VV'FVV'p. \tag{6.1.5}$$

Since the mixed estimator is the least square estimator for the augmented model (3.3.1) its conditional variance and average MSE is

$$v = p'(X'X\tau^2 + R'R\sigma^2)^+p\sigma^2\tau^2. \tag{6.1.6}$$

Example 6.1.1. MSE of Estimable and Non-Estimable Parametric Functions. In Example 4.2.1 $\alpha_1 - \alpha_2$ was an estimable parametric function but α_1 was not. The reader should show that (Exercise 6.1.1)

$$(X'X)^+ = \begin{bmatrix} \frac{1}{9} & \frac{1}{18} & \frac{1}{18} \\ \frac{1}{18} & \frac{5}{18} & -\frac{4}{18} \\ \frac{1}{18} & -\frac{4}{18} & \frac{5}{18} \end{bmatrix}.$$

Let $\delta' = [\mu \quad \alpha_1 \quad \alpha_2]$, and $C' = [1/\sqrt{3} \quad -1/\sqrt{3} \quad -1/\sqrt{3}]$.
From Equation (6.1.4) the conditional MSE of α_2 is, for $\theta = 0$,

$$
m_2 = \begin{bmatrix} 0 & 1 & 0 \end{bmatrix} \begin{bmatrix} \frac{1}{9} & \frac{1}{18} & \frac{1}{18} \\ \frac{1}{18} & \frac{5}{18} & -\frac{4}{18} \\ \frac{1}{18} & -\frac{4}{18} & \frac{5}{18} \end{bmatrix} \begin{bmatrix} 0 \\ 1 \\ 0 \end{bmatrix} \sigma^2
$$

$$
+ \begin{bmatrix} 0 & 1 & 0 \end{bmatrix} CC'\delta\delta'CC' \begin{bmatrix} 0 \\ 1 \\ 0 \end{bmatrix}
$$

$$
= \frac{5}{18}\sigma^2 + \left[-\frac{1}{3}\mu + \frac{1}{3}\alpha_1 + \frac{1}{3}\alpha_2\right]^2.
$$

For F in Example 4.2.1

$$
m_1 = \frac{5}{18}\sigma^2 + \begin{bmatrix} 0 & 1 & 0 \end{bmatrix} N \begin{bmatrix} 0 \\ 1 \\ 0 \end{bmatrix} = \frac{5}{18}\sigma^2 + \frac{2}{3}
$$

with

$$
N = \begin{bmatrix} \frac{1}{3} & -\frac{1}{3} & -\frac{1}{3} \\ -\frac{1}{3} & \frac{1}{3} & \frac{1}{3} \\ -\frac{1}{3} & \frac{1}{3} & \frac{1}{3} \end{bmatrix} \begin{bmatrix} 3 & 2 & 0 \\ 0 & 2 & 0 \\ 0 & 0 & 1 \end{bmatrix} \begin{bmatrix} \frac{1}{3} & -\frac{1}{3} & -\frac{1}{3} \\ -\frac{1}{3} & \frac{1}{3} & \frac{1}{3} \\ -\frac{1}{3} & \frac{1}{3} & \frac{1}{3} \end{bmatrix}.
$$

However, for $\alpha_1 - \alpha_2$, $p'V = 0$. Thus,

$$m_1 = m_2 = [0 \; 1 - 1] \begin{bmatrix} \frac{1}{9} & \frac{1}{18} & \frac{1}{18} \\ \frac{1}{18} & \frac{5}{18} & -\frac{4}{18} \\ \frac{1}{18} & -\frac{4}{18} & \frac{5}{18} \end{bmatrix} \begin{bmatrix} 0 \\ 1 \\ -1 \end{bmatrix} \sigma^2$$

$$= \sigma^2.$$

Exercise 6.1.1. A. Verify that $(X'X)^+$ is as stated in Example 6.1.1 above.

B. Show that if

$$G = \begin{bmatrix} 0 & 0 & 0 \\ 0 & \frac{1}{2} & 0 \\ 0 & 0 & \frac{1}{2} \end{bmatrix}$$

is used in place of $(X'X)^+$ the variance is the same for $\alpha_1 - \alpha_2$ but not for α_1.

Exercise 6.1.2. Show that the average bias of a LS estimator averaged over a prior with dispersion $F = UDU'$ is zero.

Example 6.1.2. Different Kinds of Estimability. Consider the augmented model

$$\begin{bmatrix} Y \\ r \end{bmatrix} = \begin{bmatrix} X \\ R \end{bmatrix} \beta + \begin{bmatrix} \epsilon \\ \phi \end{bmatrix},$$

where X is as in Example 4.2.1,

$$E(\epsilon) = 0, \quad D(\epsilon) = I,$$
$$E(\phi) = 0, \quad D(\phi) = I,$$

$$r = \begin{bmatrix} 0 & \frac{1}{\sqrt{2}} & \frac{1}{\sqrt{2}} \\ 0 & \frac{1}{\sqrt{2}} & -\frac{1}{\sqrt{2}} \end{bmatrix} \begin{bmatrix} \mu \\ \alpha_1 \\ \alpha_2 \end{bmatrix} + \phi.$$

Now for $\alpha_1 - \alpha_2$ from (6.1.6) (The reader may verify the computation in Exercise 6.1.3).

$$V = \begin{bmatrix} 0 & 1 & -1 \end{bmatrix} \begin{bmatrix} \frac{3}{4} & -\frac{1}{2} & -\frac{1}{2} \\ -\frac{1}{2} & \frac{2}{3} & \frac{1}{3} \\ -\frac{1}{2} & \frac{1}{3} & \frac{2}{3} \end{bmatrix} \begin{bmatrix} 0 \\ 1 \\ -1 \end{bmatrix} = \frac{2}{3}.$$

Exercise 6.1.3. A. For the model in Example 6.1.2 verify

$$(X'X + R'R)^+ = \begin{bmatrix} \frac{3}{4} & -\frac{1}{2} & -\frac{1}{2} \\ -\frac{1}{2} & \frac{2}{3} & \frac{1}{3} \\ -\frac{1}{2} & \frac{1}{3} & \frac{2}{3} \end{bmatrix}$$

and $V = 2/3$.
B. Explain why α_1 is (X, R) estimable but not X estimable.

6.2 Relationship Between the Average Variance and the MSE

It will be shown that the average MSE of the BE is the difference between the variance of the prior distribution and the variance of $p'\hat{\beta}$. In symbols

$$m_1 = p'Fp - \text{Var}(p'\hat{\beta}). \tag{6.2.1}$$

The variance of $p'\hat{\beta}_b$ is

$$v = p'FX'(XFX' + \sigma^2 I)^{-1}XFp. \tag{6.2.2}$$

The MSE of (3.5.7) may be obtained by substitution of (3.5.6) into (3.5.3). Thus,

$$m_1 = p'Fp - p'FX'(XFX' + \sigma^2 I)^{-1}XFp. \tag{6.2.3}$$

The result (6.2.1) follows.

When the data and the prior distribution are normal (6.2.3) states that the variance of the posterior distribution is the difference between the prior variance and the variance of the mean of the posterior distribution. In this case notice that

$$\text{Var}(p'\beta) = [E(\text{Var}_{\underline{x}}(p'\beta)) + \text{Var } E_{\underline{x}}(p'\beta)]. \tag{6.2.4}$$

The $\underline{x}$ represents a random sample from a population. Result (6.2.4) is really an application of the well known fact that if X and Y are random variables

$$\text{Var}(Y) = E[\text{Var}(Y|X)] + \text{Var } E(Y|X). \tag{6.2.5}$$

The MSE of the alternative forms will be obtained in the next section by first finding the variance of $p'\hat{\beta}$ and then employing (6.2.1). Considerable algebra will be saved that way.

Example 6.2.1. MSE of BE of Normal Parameter. In Example 2.2.1 the variance of the prior distribution is σ_0^2. Also $\theta^{(b)} =$

$E(\hat{\theta}|\bar{X})$. Now

$$\mathrm{Var}(\theta^{(b)}) = \mathrm{Var}\Big[\frac{\bar{X}\sigma_0^2 + \frac{\theta_0\sigma^2}{n}}{\sigma_0^2 + \frac{\sigma^2}{n}}\Big]$$

$$= \frac{\sigma_0^4}{(\sigma_0^2 + \frac{\sigma^2}{n})^2} \, \mathrm{Var}(\bar{X}).$$

From (6.2.4)

$$\mathrm{Var}(\bar{X}) = E[\,\mathrm{Var}(\bar{X}|\theta)] + \mathrm{Var}\,E[\bar{X}|\theta]$$

$$= \frac{\sigma^2}{n} + \sigma_0^2.$$

Thus,

$$\mathrm{Var}(\theta^{(b)}) = \frac{\sigma_0^4}{(\sigma_0^2 + \frac{\sigma^2}{n})}.$$

Now, from Equation (6.2.1),

$$m_1 = \sigma_0^2 - \frac{\sigma_0^4}{(\sigma_0^2 + \frac{\sigma^2}{n})} = \frac{\frac{\sigma^2}{n}\sigma_0^2}{(\sigma_0^2 + \frac{\sigma^2}{n})} = \sigma^2 - \frac{\frac{\sigma^4}{n^2}}{(\sigma_0^2 + \frac{\sigma^2}{n})}.$$

Exercise 6.2.1. Verify that for the prior assumptions on θ in Example 2.2.1

$$\mathrm{Var}(\theta) = E[\,\mathrm{Var}(\theta|\bar{X})] + \mathrm{Var}\,E(\theta|\bar{X}).$$

6.3 The Average Variance and the MSE of the BE

The average variance will be calculated for the different forms of the BE. Their average MSE will then be calculated from Equation (6.2.1).

First observe that the dispersion of Y (the matrix of variances and covariances)

$$D(Y) = XFX' + \sigma^2 I. \tag{6.3.1}$$

Thus,

$$D(b) = D[(X'X)^+ X'Y]$$

$$= (X'X)^+ X'[XFX' + \sigma^2 I]X(X'X)^+ \tag{6.3.2}$$

$$= UU'FUU' + \sigma^2 (X'X)^+.$$

Using (6.3.1) the variance of the first two alternative forms (4.2.4) and (4.2.6) of the BE is, since $V(p'\hat{\beta}) = p'D(\hat{\beta})p$,

$$\text{Var}(p'\hat{\beta}) = p'[X'X + \sigma^2(UU'FUU')^+]^+[X'XFX'X + \sigma^2 X'X] \tag{6.3.3}$$

$$\times [X'X + \sigma^2(UU'FUU')^+]^+ p.$$

Equation (6.3.3) can be simplified considerably. For this purpose, notice that

$$[X'X + \sigma^2(UU'FUU')^+]^+ X'X$$

$$= U(\Lambda + \sigma^2(U'FU)^{-1})^{-1}\Lambda U' \tag{6.3.4}$$

$$= UU' - U(\Lambda + \sigma^2(U'FU)^{-1})^{-1}(U'FU)^{-1}U'\sigma^2.$$

Thus,

$$[X'X + \sigma^2(UU'FUU')^+]X'XFX'X[X'X + \sigma^2(UU'FUU')^+]^+$$

$$= [UU' - U(\Lambda + \sigma^2(U'FU)^{-1})^{-1}(U'FU)^{-1}U'\sigma^2]F$$

$$\times [UU' - \sigma^2 U(U'FU)^{-1}(\Lambda + \sigma^2(U'FU)^{-1})^{-1}U']$$

$$=UU'FUU' - 2U(\Lambda + \sigma^2(U'FU)^{-1})^{-1}U'\sigma^2 \qquad (6.3.5)$$

$$+ U(\Lambda + \sigma^2(U'FU)^{-1})^{-1}(U'FU)^{-1}$$

$$\times (\Lambda + \sigma^2(U'FU)^{-1})^{-1}\sigma^4$$

$$=UU'FUU' - 2[X'X + \sigma^2(UU'FUU')^+]^+\sigma^2$$

$$+ [X'X + \sigma^2(UU'FUU')^+]^+(UU'FUU')^+$$

$$\times [X'X + \sigma^2(UU'FUU')^+]^+\sigma^4.$$

Substitution in (6.3.3) yields the simplified form of the variance, namely,

$$\text{Var}(p'\hat{\beta}) = p'Fp - p'(X'X + \sigma^2(UU'FUU')^+)^+ p\sigma^2. \qquad (6.3.6)$$

Equation (6.3.6) shows that the variance of $p'\hat{\beta}$ is less than the variance of $p'\beta$. Substituting (6.3.6) into (6.2.1) gives the MSE

$$m_1 = p'(X'X + \sigma^2(UU'FUU')^+)^+ p\sigma^2. \qquad (6.3.7)$$

Notice the similarity between (6.3.7) and (6.1.6). Clarification of this similarity will be given later.

The forms of the BE (4.2.7) and (4.2.8) have variance

$$\text{Var}(p'\hat{\beta}) = p'F[UU'FUU' + \sigma^2(X'X)^+]^+ Fp \qquad (6.3.8)$$

and MSE

$$m_1 = p'Fp - p'F[UU'FUU' + \sigma^2(X'X)^+]^+ Fp. \qquad (6.3.9)$$

Equation (6.3.9) shows that the MSE is less than the prior variance.

The forms of the BE (4.2.1) have variance

$$\text{Var } (p'\hat{\beta}) = p'Fp - p'(X'X)^+p\sigma^2 \tag{6.3.10}$$

$$+ p'(X'X)^+[UU'FUU' + \sigma^2(X'X)^+]^+(X'X)^+p\sigma^4.$$

Its

$$\text{MSE} = p'(X'X)^+p\sigma^2$$

$$\tag{6.3.11}$$

$$- p'(X'X)^+[UU'FUU' + \sigma^2(X'X)^+]^+(X'X)^+p\sigma^4.$$

The second term in the MSE is the improvement over the MSE of the LS estimator.

Example 6.3.1. Average MSE of a Contraction Estimator.
For the prior where

$$E(\beta) = 0 \quad \text{and} \quad D(\beta) = \omega^2(X'X)^+,$$

the BE takes the form of the contraction estimator of Mayer and Wilke. Then

$$p'\hat{\beta} = \frac{\omega^2}{\sigma^2 + \omega^2}p'b.$$

Now

$$\text{Var}(p'\hat{\beta}) = \frac{\omega^4}{\sigma^2 + \omega^2}p'(X'X)^+p.$$

From Equation (6.2.1)

$$m_1 = p'(X'X)^+ p\omega^2 - \frac{\omega^4}{\sigma^2 + \omega^2} p'(X'X)^+ p$$

$$= \frac{\sigma^2 \omega^2}{\sigma^2 + \omega^2} p'(X'X)^+ p$$

$$= \omega^2 p'(X'X)^+ p - \frac{\omega^4}{\sigma^2 + \omega^2} p'(X'X)^+ p$$

$$= \sigma^2 p'(X'X)^+ p - \frac{\sigma^4}{\sigma^2 + \omega^2} p'(X'X)^+ p.$$

Exercise 6.3.1. Exhibit the three alternative forms of the MSE of the ordinary ridge, generalized ridge and generalized contraction estimator.

Exercise 6.3.2. For a linear model with

$$X = \begin{bmatrix} 1_3 & 1_3 & 0 \\ 1_3 & 0 & 1_3 \end{bmatrix}, \quad F = \begin{bmatrix} 3 & 0 & 0 \\ 0 & 2 & 0 \\ 0 & 0 & 0 \end{bmatrix},$$

$\sigma^2 = 1$, what is the average MSE of $\alpha_1^{(b)} - \alpha_2^{(b)}$?

Excersise 6.3.3. Let $p'\beta$ be an estimable parametric function. Assume that the prior is such that $U'FU$ is positive definite. Under what conditions are the MSE expressions (6.3.7), (6.3.9), and (6.3.11) independent of the choice of generalized inverse? In other words when can $(X'X)^+$ be replaced by any generalized inverse G? Prove your assertions.

6.4 Alternative Forms of the MSE of the Mixed Estimator

In Chapter IV it was shown how the BE could be looked upon as:

1. a BE with respect to the augmented model (3.3.1) with unknown mean and dispersion F;
2. a BE with respect to the linear model (1.2.1) with prior mean b_2 and variance $\tau^2(R'R)^+$;
3. a least square estimator with respect to the augmented model (3.3.1).

Thus the mathematical form of the variance of the mixed estimator as viewed in 3 above is the same as the average MSE in 1 and 2.

The forms of the MSE corresponding to (6.3.7), (6.3.9) and (6.3.11) are

$$m(\hat{\beta})$$

$$= p'(X'X\tau^2 + R'R\sigma^2)^+ p\sigma^2\tau^2 \qquad (6.4.1)$$

$$= p'(X'X)^+ p\sigma^2 - p'(X'X)^+[\sigma^2(X'X)^+ + \tau^2(R'R)^+]^+(X'X)^+ p\sigma^4 \qquad (6.4.2)$$

$$= p'(R'R)^+ p\tau^2 - p'(R'R)^+[\sigma^2(X'X)^+ + \tau^2(R'R)^+]^+(R'R)^+ p\tau^4. \qquad (6.4.3)$$

Notice that:

1. The above equations are the same as those for the BE with the matrix $(R'R)^+\tau^2$ in place of $(UU'FUU')^+$.
2. The MSE is less than that of b_1 and b_2.

Also note the similarity between (6.4.1) and (6.3.7). Equations (6.4.2) and (6.4.3) are the same with the roles of sample and prior information exchanged.

Example 6.4.1. MSE of Mixed Estimator. Consider the augmented model

$$Y_i = \mu + \epsilon_i, \quad 1 \le i \le m,$$
$$r_j = \mu + \phi_j, \quad 1 \le j \le n$$

Assume

$$E(\epsilon_i) = 0, \quad D(\epsilon_i) = \sigma^2,$$
$$E(\phi_j) = 0, \quad D(\phi_j) = \tau^2.$$

Then

$$\hat{\mu} = \frac{1}{m\tau^2 + n\sigma^2} \left(\frac{m}{\tau^2} \bar{Y} + \frac{n}{\sigma^2} \bar{r} \right).$$

This estimator has MSE by (6.4.1)

$$m_1 = \frac{\sigma^2 \tau^2}{m\tau^2 + n\sigma^2} = \frac{\frac{\sigma^2 \tau^2}{mn}}{\frac{\tau^2}{n} + \frac{\sigma^2}{m}}$$

$$= \frac{\frac{\sigma^2}{m} \left(\frac{\tau^2}{n} + \frac{\sigma^2}{m} \right)}{\frac{\tau^2}{n} + \frac{\sigma^2}{m}} - \frac{\frac{\sigma^4}{m^2}}{\frac{\tau^2}{n} + \frac{\sigma^2}{m}}$$

$$= \frac{\sigma^2}{m} - \frac{\sigma^2}{m} \cdot \frac{1}{\left(\frac{\tau^2}{n} + \frac{\sigma^2}{m} \right)} \cdot \frac{\sigma^2}{m}.$$

Similarly

$$m_1 = \frac{\tau^2}{n} - \frac{\tau^2}{n} \cdot \frac{1}{\left(\frac{\tau^2}{n} + \frac{\sigma^2}{m} \right)} \cdot \frac{\tau^2}{n}. \tag{6.4.4}$$

Exercise 6.4.1. Verify (6.4.4).

Exercise 6.4.2. Consider the model above except

$$r_j = c\mu + \phi_j, \quad 1 \le j \le m.$$

Now find the alternative forms of the MSE of the mixed estimator.

Exercise 6.4.3. For the models

$$
Y = \begin{bmatrix} 1_3 & 1_3 & 0 \\ 1_3 & 0 & 1_3 \end{bmatrix} \begin{bmatrix} \mu \\ \alpha_1 \\ \alpha_2 \end{bmatrix} + \epsilon, \qquad \text{(Model I)}
$$

$$
r = \begin{bmatrix} -\frac{2}{\sqrt{18}} & \frac{1}{\sqrt{18}} & \frac{1}{\sqrt{18}} \\ 0 & -\frac{1}{2} & \frac{1}{2} \end{bmatrix} \begin{bmatrix} \mu \\ \alpha_1 \\ \alpha_2 \end{bmatrix} + \epsilon : \qquad \text{(Model II)}
$$

A. Find the alternative forms of the MSE of the mixed estimator.
B. Show that $\alpha_1 - \alpha_2$ is an X and R estimable parametric function.
C. Find the MSE of both the LS and the mixed estimator for $\alpha_1 - \alpha_2$.

6.5 Comparison of the MSE of Different BE

The following theorem shows that for a given value of the prior mean, the MSE of a BE becomes smaller as the prior mean becomes more precise. It will be used to obtain inequalities on the prior parameters to compare various ridge type estimators in Section 6.6.

Theorem 6.5.1. *Suppose that $F_2 - F_1$ is non-negative definite. Let $p'\hat{\beta}_1$ be the BE associated with prior mean θ_1 and dispersion F_1. Let $p'\hat{\beta}_2$ be the BE associated with prior mean θ_2 and dispersion F_2. Then*

$$
\text{MSE}(p'\hat{\beta}_1) \leq \text{MSE}(p'\hat{\beta}_2). \qquad (6.5.1)
$$

Proof. Recall that $A \leq B$ if $B - A$ is non-negative definite. Thus, $F_1 \leq F_2$ implies

$$
U'F_1U + \sigma^2\Lambda^{-1} \leq U'F_2U + \sigma^2\Lambda^{-1}. \qquad (6.5.2)
$$

From C.R. Rao (1975)

$$(U'F_2U + \sigma^2\Lambda^{-1})^{-1} \le (U'F_1U + \sigma^2\Lambda^{-1})^{-1}. \qquad (6.5.3)$$

Thus,

$$\Lambda^{-1}(U'F_2U + \sigma^2\Lambda^{-1})^{-1}\Lambda^{-1}\sigma^4 \le \Lambda^{-1}(U'F_1U + \sigma^2\Lambda^{-1})^{-1}\Lambda^{-1}\sigma^4$$
$$(6.5.4)$$

and

$$p'(X'X)^+(UU'F_2UU' + \sigma^2(X'X)^+)^+(X'X)^+p\sigma^4$$
$$(6.5.5)$$
$$\le p'(X'X)^+(UU'F_1UU' + \sigma^2(X'X)^+)^+(X'X)^+p\sigma^4.$$

Consequently

$$\mathrm{MSE}(p'\hat{\beta}_1)$$

$$= p'(X'X)^+p\sigma^2$$

$$- p'(X'X)^+(UU'F_1UU' + \sigma^2(X'X)^+)^+(X'X)^+p\sigma^4$$

$$\le p'(X'X)^+p\sigma^2$$

$$- p'(X'X)^+(UU'F_2UU' + \sigma^2(X'X)^+)^+(X^2X)^+p\sigma^4$$

$$= \mathrm{MSE}(p'\hat{\beta}_2). \blacksquare \qquad (6.5.6)$$

Thus, for any prior mean vector a smaller prior dispersion vector means that the linear BE has a smaller MSE or the posterior distribution has a smaller variance. Numerical examples of Theorem 6.5.1 will now be given.

Example 6.5.1. Illustration of Theorem 6.5.1. Consider the linear model

$$Y = \begin{bmatrix} 1 & 1 & 0 \\ 1 & 1 & 0 \\ 1 & 0 & 1 \\ 1 & 0 & 1 \end{bmatrix} \begin{bmatrix} \mu \\ \alpha_1 \\ \alpha_2 \end{bmatrix} + \epsilon,$$

where $E(\epsilon) = 0$ and $D(\epsilon) = I$. Let θ_1 and θ_2 be the prior means. Assume

$$F_1 = \begin{bmatrix} 2 & 0 & 0 \\ 0 & 1 & 0 \\ 0 & 0 & 0 \end{bmatrix} \quad \text{and} \quad F_2 = \begin{bmatrix} 4 & 0 & 0 \\ 0 & 2 & 0 \\ 0 & 0 & 0 \end{bmatrix}.$$

Then

$$F_2 - F_1 = \begin{bmatrix} 2 & 0 & 0 \\ 0 & 1 & 0 \\ 0 & 0 & 0 \end{bmatrix}$$

is NND. Now

$$X'X = \begin{bmatrix} \frac{2}{\sqrt{6}} & 0 \\ \frac{1}{\sqrt{6}} & -\frac{1}{\sqrt{2}} \\ \frac{1}{\sqrt{6}} & \frac{1}{\sqrt{2}} \end{bmatrix} \begin{bmatrix} 6 & 0 \\ 0 & 2 \end{bmatrix} \begin{bmatrix} \frac{2}{\sqrt{6}} & \frac{1}{\sqrt{6}} & \frac{1}{\sqrt{6}} \\ 0 & -\frac{1}{\sqrt{2}} & \frac{1}{\sqrt{2}} \end{bmatrix}$$

A comparison of the MSE of $\alpha_1^{(b)} - \alpha_2^{(b)}$ will be made for the two prior dispersions. First find d' so $p' = d'U'$. Thus solve

$$\begin{bmatrix} 0 \\ 1 \\ -1 \end{bmatrix} = [d_1 \quad d_2] \begin{bmatrix} \frac{2}{\sqrt{6}} & \frac{1}{\sqrt{6}} & \frac{1}{\sqrt{6}} \\ 0 & -\frac{1}{\sqrt{2}} & \frac{1}{\sqrt{2}} \end{bmatrix}$$

or

$$\frac{2}{\sqrt{6}} d_1 = 0,$$

$$\frac{1}{\sqrt{6}} d_1 - \frac{1}{\sqrt{2}} d_2 = 1,$$

$$\frac{1}{\sqrt{6}} d_1 + \frac{1}{\sqrt{2}} d_2 = -1$$

obtaining $d_1 = 0$, $d_2 = -\sqrt{2}$. Now for estimable parametric functions

$$m_1 = d'(\Lambda + \sigma^2 (U'FU)^{-1})^{-1} d\sigma^2.$$

For F_1,

$$U'F_1 U = \begin{bmatrix} \frac{2}{\sqrt{6}} & \frac{1}{\sqrt{6}} & \frac{1}{\sqrt{6}} \\ 0 & -\frac{1}{\sqrt{2}} & \frac{1}{\sqrt{2}} \end{bmatrix} \begin{bmatrix} 2 & 0 & 0 \\ 0 & 1 & 0 \\ 0 & 0 & 0 \end{bmatrix} \begin{bmatrix} \frac{2}{\sqrt{6}} & 0 \\ \frac{1}{\sqrt{6}} & -\frac{1}{\sqrt{2}} \\ \frac{1}{\sqrt{6}} & \frac{1}{\sqrt{2}} \end{bmatrix}$$

$$= \begin{bmatrix} \frac{3}{2} & -\frac{1}{\sqrt{12}} \\ -\frac{1}{\sqrt{12}} & \frac{1}{2} \end{bmatrix},$$

$$(U'F_1 U)^{-1} = \frac{3}{2} \begin{bmatrix} \frac{1}{2} & \frac{1}{\sqrt{12}} \\ \frac{1}{\sqrt{12}} & \frac{3}{2} \end{bmatrix},$$

and

$$(U'F_1 U)^{-1} + \Lambda = \begin{bmatrix} \frac{27}{4} & \frac{1}{\sqrt{12}} \\ \frac{1}{\sqrt{12}} & \frac{17}{4} \end{bmatrix}.$$

Then

$$[(U'F_1U)^{-1} + \Lambda]^{-1} = \frac{1}{28.604} \begin{bmatrix} \frac{17}{4} & -\frac{1}{\sqrt{12}} \\ -\frac{1}{\sqrt{12}} & \frac{27}{4} \end{bmatrix}.$$

For the particular choice of d,

$$m_1 = \frac{2}{28.604}\left(\frac{27}{4}\right) = .4720.$$

For F_2,

$$U'F_2U = \begin{bmatrix} 2 & -\frac{2}{\sqrt{12}} \\ -\frac{2}{\sqrt{12}} & 1 \end{bmatrix},$$

$$(U'F_2U)^{-1} = \frac{3}{5} \begin{bmatrix} 1 & \frac{2}{\sqrt{12}} \\ \frac{2}{\sqrt{12}} & 2 \end{bmatrix},$$

$$(U'F_2U)^{-1} + \Lambda = \begin{bmatrix} 7 & \frac{2}{\sqrt{12}} \\ \frac{2}{\sqrt{12}} & 4 \end{bmatrix}$$

and

$$[(U'F_2U)^{-1} + \Lambda]^{-1} = \frac{3}{83} \begin{bmatrix} 4 & -\frac{2}{\sqrt{12}} \\ -\frac{2}{\sqrt{12}} & 7 \end{bmatrix}.$$

Then

$$m_2 = 2\left(\frac{3}{83}\right)(7) = .506.$$

Thus $m_2 > m_1$ as expected.

Exercise 6.5.1. For Example 6.5.1 above show that the MSE for $\hat{\mu}^{(b)} + \alpha_2^{(b)}$ is smaller for F_1 than for F_2.

Exercise 6.5.2. Redo Example 6.5.1 where

$$F_1 = \begin{bmatrix} 2 & 1 & 0 \\ 1 & 2 & 0 \\ 0 & 0 & 0 \end{bmatrix} \quad \text{and} \quad F_2 = \begin{bmatrix} 4 & 1 & 0 \\ 1 & 2 & 0 \\ 0 & 0 & 0 \end{bmatrix}.$$

Exercise 6.5.3. Consider a linear model where

$$X = \begin{bmatrix} 3 & -1 & -1 & -1 \\ -1 & \frac{7}{3} & -\frac{2}{3} & -\frac{2}{3} \\ -1 & -\frac{2}{3} & \frac{11}{6} & -\frac{1}{6} \\ -1 & -\frac{2}{3} & -\frac{1}{6} & \frac{11}{6} \end{bmatrix}, \quad \beta = \begin{bmatrix} \beta_1 \\ \beta_2 \\ \beta_3 \\ \beta_4 \end{bmatrix} \quad \text{and} \quad \sigma^2 = 2.$$

Assume the parameters of the model have prior distributions with

$$E(\beta_1) = \begin{bmatrix} -1 \\ 4 \\ 2 \\ 0 \end{bmatrix}, \quad D(\beta_1) = \begin{bmatrix} 2 & -1 & -1 & 0 \\ 0 & 2 & -1 & -1 \\ 0 & -1 & 2 & -1 \\ 0 & -1 & -1 & 2 \end{bmatrix},$$

and

$$E(\beta_2) = \begin{bmatrix} 2 \\ -1 \\ -1 \\ 1 \end{bmatrix}, \quad D(\beta_2) = \begin{bmatrix} 3 & -1 & -1 & -1 \\ -1 & 3 & -1 & -1 \\ -1 & -1 & 3 & -1 \\ -1 & -1 & -1 & 3 \end{bmatrix}.$$

Show by explicit computation that the BE of

$$\beta_1 + \frac{2}{3}\beta_2 + \frac{1}{6}\beta_3 - \frac{11}{6}\beta_4$$

has a smaller MSE for the first prior than for the second prior.

A corresponding result for mixed estimators is:

Theorem 6.5.2. *Consider the augmented model*

$$\begin{bmatrix} Y \\ r \end{bmatrix} = \begin{bmatrix} X \\ R \end{bmatrix} \beta + \begin{bmatrix} \epsilon \\ \phi \end{bmatrix}. \tag{6.5.7}$$

Let $H_2 - H_1$ *be non-negative definite and* H_1 *and* H_2 *be positive definite. Let* $\hat{\beta}_{1e}$ *be the mixed estimator derived from the assumption*

$$E(\phi) = 0 \quad \text{and} \quad D(\phi) = H_1, \tag{6.5.8}$$

$\hat{\beta}_{2e}$ *be the mixed estimator derived from the assumptions*

$$E(\phi) = 0 \quad \text{and} \quad D(\phi) = H_2. \tag{6.5.9}$$

Then

$$\text{Var}(p'\hat{\beta}_{1e}) \le \text{Var}(p'\hat{\beta}_{2e}). \tag{6.5.10}$$

The proof is similar to that of Theorem 6.5.1.

In other words, a smaller dispersion of the error term in the stochastic prior assumptions means a smaller variance for the mixed estimator, the least square estimator or the augmented model.

Exercise 6.5.4. In Theorem 6.5.2 let

$$X = \begin{bmatrix} 1 & 1 & 0 \\ \frac{1}{\sqrt{2}} & \frac{1}{\sqrt{2}} & 0 \\ 0 & 0 & 1 \end{bmatrix}, \quad R = \begin{bmatrix} 1 & 0 & 0 \\ 0 & 1 & 0 \\ 0 & 0 & 1 \end{bmatrix},$$

$$D(\epsilon) = I_3,$$

$$H_1 = \begin{bmatrix} \frac{2}{3} & 0 & 0 \\ 0 & 2 & 0 \\ 0 & 0 & 1 \end{bmatrix} \quad \text{and} \quad H_2 = \begin{bmatrix} 1 & 0 & 0 \\ 0 & 2 & 0 \\ 0 & 0 & 1 \end{bmatrix}.$$

A. Find the mixed estimator of

$$\beta_0 + \beta_1 + \frac{1}{\sqrt{2}}\beta_2$$

for both H_1 and H_2.

B. Find the variance of these two estimators and compare. Are the results of the comparison the same as those obtained by Theorem 6.5.2?

Exercise 6.5.5. Prove Theorem 6.5.2.

6.6 Comparison of the Ridge and Contraction Estimator's MSE

Explicit expressions for these estimators are presented in Section 4.5. From the main result of Section 6.5 the following conclusions are arrived at regarding their relative merits. The ith latent root of $X'X$ (eigenvalue) is denoted by λ_i and the ith element of K and C are denoted by k_i and c_i.

1. The contraction estimator has smaller average MSE than the ridge estimator if $k/c < \lambda_i$.
2. The generalized contraction estimator has a smaller average MSE than the generalized ridge estimater if $k_i/c_i < \lambda_i$.
3. The generalized ridge estimator is superior to the ridge estimator provided $k_i/k > 1$.
4. The generalized contraction estimator is superior to the contraction estimator provided $k_i/k > 1$.
5. The contraction estimator has a smaller average MSE than the generalized ridge estimator provided $k_i/c < \lambda_i$.
6. The ridge estimator is superior to the generalized contraction estimator provided $k/c_i > \lambda_i$.

Example 6.6.1. Comparison of MSE of Different Ridge and Contraction Estimators. For the data in Example 1.2.1 :

1. The contraction estimator has a smaller average MSE than the generalized ridge regression estimator when $k/c < .025724$.
2. The generalized contraction estimator has a smaller average MSE than the generalized ridge estmator if

$$k_1 < (1.5561 \times 10^{11})c_1,$$
$$k_2 < (1.8451 \times 10^{7})c_2,$$
$$k_3 < 3989.7c_3,$$
$$k_4 < 0.025724c_4.$$

(Notice that k_1 and k_2 have a much wider range of permissible values.)
3. The ridge regression estimator is superior to the generalized contraction estimator, i.e., it has a smaller MSE provided

$$k > (1.5561 \times 10^{11})c_1,$$
$$k > (1.8451 \times 10^{7})c_2,$$
$$k > 3989.7c_3,$$
$$k > 0.025724c_4.$$

Exercise 6.6.1. Give criteria for comparison of the various ridge and contraction estimators for the X matrix in Example 6.5.1.

Exercise 6.6.2. Let X be as in Exercise 6.5.3 above for the cases $c = 2$ and $c = 1/2$. Find an appropriate upper bound on k where the contraction estimator is superior to the ridge estimator.

Similar statements may also be made about the variances of $p'\hat{\beta}$ provided that $F = UDU'$ where D is a diagonal matrix. Let d_i represent the elements of D. The variance v in (6.3.6) may be written

$$v = p'U[D - (\Lambda + \sigma^2 D^{-1})\sigma^2]U'p = p'U\Theta U', \qquad (6.6.1)$$

where $\Theta = D - (\Lambda + \sigma^2 D^{-1})^{-1}\sigma^2$. The elements of Θ are

$$\theta_i = d_i - \frac{\sigma^2}{\lambda_i + \frac{\sigma^2}{d_i}} = \frac{\sigma^2 d_i}{\lambda_i d_i + \sigma^2} = g(d_i). \qquad (6.6.2)$$

Now $g(d_i)$ is an increasing function because

$$g'(d_i) = \frac{\sigma^4}{(\lambda_i d_i + \sigma^2)^2} > 0. \qquad (6.6.3)$$

Thus, the variance of $p'\hat{\beta}$ becomes smaller as the prior dispersion becomes smaller.

Example 6.6.2. A Counterexample. Unfortunately, when F is not of the form UDU' the same comparison of the dispersions cannot be made. Here is a counterexample. Let $X'X = I$, $\sigma^2 = 1$. Let $m = 2$,

$$F_1 = \begin{bmatrix} 2 & 1 \\ 1 & 1 \end{bmatrix}, \quad F_2 = \begin{bmatrix} 2 & 1 \\ 1 & 3 \end{bmatrix}.$$

Then $F_2 - F_1 = \begin{bmatrix} 0 & 0 \\ 0 & 2 \end{bmatrix}$, a NND matrix. Now

$$F - (I + F^{-1})^{-1} = F(F + I)^{-1}F.$$

Also

$$F_1(F_1 + I)^{-1}F_1 = \begin{bmatrix} \frac{7}{5} & \frac{4}{5} \\ \frac{4}{5} & \frac{3}{5} \end{bmatrix},$$

$$F_2(F_2 + I)^{-1}F_2 = \begin{bmatrix} \frac{15}{11} & \frac{10}{11} \\ \frac{10}{11} & \frac{25}{11} \end{bmatrix},$$

and

$$F_2(F_2 + I)^{-1}F_2 - F_1(F_1 + I)^{-1}F_1 = \begin{bmatrix} -\frac{2}{55} & \frac{6}{55} \\ \frac{6}{55} & \frac{92}{55} \end{bmatrix}.$$

The above matrix is not NND.

Exercise 6.6.3. In Example 6.6.2 above replace F_2 by

$$F_2 = \begin{bmatrix} 2 & 1 \\ 1 & 2 + \epsilon \end{bmatrix},$$

where $\epsilon > 0$. Show that the difference of the average dispersions is not NND.

6.7 Summary and Remarks

In this chapter:

1. The mathematical form of the measure of goodness for the BE, mixed and minimax estimators were obtained.
2. It was shown how a smaller prior dispersion led to a BE with a smaller MSE.
3. Criteria for comparison of the average MSE of different kinds of ridge estimators were given.

Each of the estimators, for the given prior assumptions, with respect to its measure of efficiency is better than the least square estimator. Also the conditions on the prior assumptions for one ridge type estimator to be better than another are particularly simple.

When the prior assumptions are neglected the Bayes estimators are better than the least square estimators for certain values of the β parameters. The problem of comparison of the goodness of two Bayes estimators neglecting the prior assumptions is more complicated.

The comparison of estimators when prior assumptions are neglected will now be taken up.

Chapter VII

The MSE Neglecting the Prior Assumptions

7.0 Introduction

The MSE neglecting prior assumptions has properties quite different from the average MSE. The average MSE of the BE is smaller than that of the least square estimator. One BE has a smaller average MSE than another if the prior information is more precise. The MSE of the BE or a mixed estimator neglecting the prior observations is smaller than that of the least square estimator provided that the β parameters lie on an ellipsoid about the prior mean. A more precise prior mean does not automatically imply one estimator has a smaller MSE neglecting prior assumptions.

To deal with these problems in Section 7.1 three alternative forms of the MSE of the BE without averaging over the prior are calculated. The corresponding results for the mixed estimator are found in Section 7.2. The form of the ellipsoids where the conditional MSE of the BE is smaller than that of the LS and the average MSE is given in Section 7.3. Comparisons of ridge type estimators turn out to be special cases. Section 7.4 has the corresponding results for mixed estimators.

The comparison of the MSE of two BE is a harder problem than comparison of the MSE of a BE and a LS. The region where one

185

BE has a smaller MSE than another is more complicated. Easier comparisons are made when the MSE is averaged over a quadratic loss function. This is done in Section 7.5.

Throughout the chapter the reader should notice the similarity in the results for the BE and mixed estimator. The Bayes and the mixed estimator are solutions to different optimization problems employing different kinds of prior information.

7.1 The MSE of the BE

The MSE without averaging over the prior distribution shall now be found for the different alternative forms of the BE. As was explained in Chapter V this MSE is the sum of the conditional variance and the squared bias. Three forms of the MSE will be found. These correspond to (6.3.7), (6.3.9) and (6.3.11). In fact if the forms of the MSE below are averaged over the prior distribution the result is the average MSE computed in Section 6.3.

Let

$$C_1 = UU'FUU' + \sigma^2(X'X)^+$$

and

$$\Gamma_1 = X'X + \sigma^2(UU'FUU')^+.$$

The conditional variance of (4.2.4) and (4.2.6) is

$$\text{Var}(p'\hat{\beta}) = p'\Gamma_1^+ X'X\Gamma_1^+ p\sigma^2. \tag{7.1.1}$$

To obtain the expression for the bias first observe that

$$E(p'\hat{\beta}) = p'\theta + p'\Gamma_1^+ X'X(\beta - \theta). \tag{7.1.2}$$

Then the bias is

$$p'\beta - E(p'\hat{\beta}) = p'[I - \Gamma_1^+ X'X](\beta - \theta). \tag{7.1.3}$$

To simplify (7.1.3) notice that, since $UU' + VV' = I$,

$$I - [X'X + \sigma^2 (UU'FUU')^+]^+ X'X$$

$$= UU' + VV' - U[(\Lambda + \sigma^2 (U'FU)^{-1})^{-1}\Lambda]U'$$

$$= U(\Lambda + \sigma^2 (U'FU)^{-1})^{-1}(U'FU)^{-1}U'\sigma^2 + VV'$$

$$= (X'X + \sigma^2 (UU'FUU')^+)^+ (UU'FUU')^+ \sigma^2 + VV'. \tag{7.1.4}$$

$$= \Gamma_1^+ (UU'FUU')^+ \sigma^2 + VV'.$$

Thus, from (7.1.3) and (7.1.4), when $p'\beta$ is estimable the bias of $p'\hat{\beta}$ is

$$p'(\beta - E(\hat{\beta})) = p'\Gamma_1^+ (UU'FUU')^+(\beta - \theta). \tag{7.1.5}$$

The expression for the MSE of (4.2.4) and (4.2.6) is the sum of the variance and the squared bias. Thus, from (7.1.1) and (7.1.5), the conditional MSE

$$m$$
$$= p'\Gamma_1^+ [\sigma^2 X'X + (UU'FUU')^+(\beta - \theta)(\beta - \theta)'(UU'FUU')^+ \sigma^4]\Gamma_1^+ p. \tag{7.1.6}$$

The form of the average MSE (6.3.7) may be obtained by calculating the expectation of (7.1.6) averaged over the prior distribution.

The conditional variance of (4.2.7) and (4.2.8) is

$$\text{Var}(p'\hat{\beta}) = p'FC_1^+ X'XC_1^+ Fp\sigma^2. \tag{7.1.7}$$

The bias when $p'\beta$ is estimable is

$$p'(\beta - E(\beta)) = p'FC_1^+(\beta - \theta). \tag{7.1.8}$$

The conditional MSE is

$$m = p'FC_1^+ X'XC_1^+ Fp\sigma^2 + p'[I - FC_1^+](\beta - \theta)(\beta - \theta)'[I - C_1^+ F]p.$$
(7.1.9)

The average MSE (6.3.9) may be obtained by finding the average expectation of (7.1.9).

The variance of (4.2.10) is

$$\text{Var}(p'\hat{\beta}) = p'(X'X)^+ p\sigma^2 - 2p'(X'X)^+ C_1^+ (X'X)^+ p\sigma^4$$

$$+ p'(X'X)^+ C_1^+ (X'X)^+ C_1^+ (X'X)^+ p\sigma^6.$$
(7.1.10)

The bias of (4.2.10) is

$$p'(\beta - E(\beta)) = p'(X'X)^+ C_1^+ (\beta - \theta)\sigma^2.$$
(7.1.11)

Thus, the form of the conditional MSE is

$$\text{MSE} = p'(X'X)^+ p\sigma^2 - 2p'(X'X)^+ C_1^+ (X'X)^+ p\sigma^4$$

$$+ p'(X'X)^+ C_1^+ [(X'X)^+ \sigma^2 + (\beta - \theta)(\beta - \theta)'] C_1^+ (X'X)^+ p\sigma^4.$$
(7.1.12)

The average of (7.1.12) is (6.3.11).

The forms of the MSE might suggest that they are smaller than the least square estimator when β and θ are close. The level of closeness required will be made precise in Section 7.3.

Example 7.1.1. Conditional MSE of the Contraction Estimator. For the contraction estimator (4.5.8) $\theta = 0$ and $F = (\sigma^2/c)(X'X)^+$. Then

$$P_1 = X'X + \sigma^2 (UU'(X'X)^+ UU' \frac{\sigma^2}{c})^+$$

$$= X'X + cX'X$$

$$= (1 + c)X'X.$$
(7.1.13)

Substituting (7.1.13) in (7.1.6) the conditional MSE is

$$m = p'(X'X)^+[\sigma^2 X'X + c^2(X'X)\beta\beta'(X'X)](X'X)^+p\frac{1}{(1+c)^2}.$$
$$(7.1.14)$$

For estimable parametric functions (7.1.14) may be written in the form

$$m = (\sigma^2 p'(X'X)^+p + c^2 p'\beta\beta'p)\frac{1}{(1+c)^2}. \qquad (7.1.15)$$

Exercise 7.1.1. Show that from (7.1.15) the average MSE is

$$m = \frac{\sigma^2}{(1+c)}p'(X'X)^+p$$

$$= \frac{\sigma^2}{c}p'(X'X)^+p - \frac{\sigma^2}{c(c+1)}p'(X'X)^+p$$

$$= \sigma^2 p'(X'X)^+p - \frac{1}{c+1}\sigma^2 p'(X'X)^+p.$$

Exercise 7.1.2. Obtain the forms (7.1.9) and (7.1.12) of the conditional MSE of the contraction estimator.

Exercise 7.1.3. Obtain the forms (7.1.6), (7.1.9) and (7.1.12) of the conditional MSE for the ordinary ridge estimator, the generalized contraction estimator and the generalized ridge estimator. Then find the average MSE.

Exercise 7.1.4. Consider the model

$$Y_i = \mu + \epsilon_i, \quad 1 \le i \le n.$$

A. Show that the ridge estimator is

$$\hat{\mu}^{(n)} = \frac{n\bar{Y}}{n+k}, \quad k > 0.$$

B. Obtain its variance and its bias.
C. Let $\hat{\mu}^{(n-1)}$ be the ridge estimator obtained from the first $n-1$ observations. Let

$$\hat{\mu} = n\hat{\mu}^{(n)} - (n-1)\hat{\mu}^{(n-1)}.$$

Find the variance, bias and MSE. How do these compare with $\hat{\mu}^{(n)}$?

D. Let $\hat{\mu}_i^{(n-1)}$ be the ridge estimator obtained when the ith observation is deleted. Let

$$\hat{\mu} = \hat{\mu}^{(n)} - \frac{(n-1)}{n} \sum_{i=1}^{n} \hat{\mu}_i^{(n-1)}.$$

Find the variance and bias of this estimator and compare with $\hat{\mu}^{(n)}$.

(This is an example of the Jackknife Procedure applied to a ridge type estimator.) (See Nyquist(1988).)

7.2 The MSE of the Mixed Estimators Neglecting the Prior Assumptions

The MSE of the BE was calculated in Section 7.1 without averaging over the prior distribution. What is the analogous computation for the mixed estimator? The prior assumptions that led to the BE were that β was a random variable with mean θ and dispersion F. For the mixed estimator it was assumed that β was a constant and additional observations were taken using the model $r = R\beta + \phi$ where ϕ had mean 0 and variance $\tau^2 I$. Similar estimators resulted.

Based on the above discussion it seems reasonable to calculate the MSE neglecting the additional observations. This amounts to a calculation of the MSE conditional on random variable r, the additional observations. In this computation the estimators are no longer unbiased.

Notice that these three forms are the same as those obtained for the MSE in Section 7.1 with $(R'R)^+$ in place of $UU'FUU'$. The next step in this development is to compare the MSE of Bayes and mixed estimators with the least square estimators.

The three forms of the bias, variance and the conditional MSE corresponding to (4.4.13), (4.4.15) and (4.4.16) are given below. Let

$$\Gamma_2 = \tau^2 X'X + \sigma^2 R'R,$$
$$C_2 = (X'X)^+ \sigma^2 + \tau^2 (R'R)^+.$$

Then the three forms are

1. $\text{Bias}(p'\hat{\beta}) = p'(\beta - E(\hat{\beta})) = p'\Gamma_2^+ R'(R\beta - r)\sigma^2,$ \hfill (7.2.1)

$\text{Var}(p'\hat{\beta}) = p'\Gamma_2^+ (X'X)\Gamma_2^+ p\sigma^2\tau^4,$ \hfill (7.2.2)

$\text{MSE}(p'\hat{\beta}) = p'\Gamma_2^+ [\sigma^2\tau^4 X'X + \sigma^4 R'(R\beta - r)(R\beta - r)'R]\Gamma_2^+ p.$ \hfill (7.2.3)

2. $\text{Bias}(p'\hat{\beta}) = p'(X'X)^+ C_2^+ (\beta - b_2)\tau^2,$ \hfill (7.2.4)

$\text{Var}(p'\hat{\beta}) = p'(R'R)^+ C_2^+ (X'X)^+ C_2^+ (R'R)^+ p\sigma^2\tau^4,$ \hfill (7.2.5)

$\text{MSE}(p'\hat{\beta})$
$$= p'(R'R)^+ C_2^+ [(X'X)^+ \sigma^2 + (\beta - b_2)(\beta - b_2)']$$
$$\times C_2^+ (R'R)^+ p. \hfill (7.2.6)$$

3. $\text{Bias} = \sigma^2 p'(X'X)^+ C_2^+ (\beta - b_2),$ \hfill (7.2.7)

$\text{Var}(p'\hat{\beta}) = p'(X'X)^+ p\sigma^2 - 2p'(X'X)^+ C_2^+ (X'X)^+ p\sigma^4$
$$+ p'(X'X)^+ C_2^+ (X'X)^+ C_2^+ (X'X)^+ p\sigma^6, \hfill (7.2.8)$$

$$\begin{aligned} \text{MSE} = \; & p'(X'X)^+ p\sigma^2 - 2p'(X'X)^+ C_2^+ (X'X)^+ p\sigma^4 \\ & + p'(X'X)^+ C_2^+ [(X'X)^+ \sigma^2 + (\beta - b_2)(\beta - b_2)'] \\ & \times C_2^+ (X'X)^+ p\sigma^4. \end{aligned} \tag{7.2.9}$$

Example 7.2.1. Conditional MSE of a Mixed Estimator.
Consider the augmented linear model

$$\begin{bmatrix} Y_1 \\ Y_2 \\ Y_3 \end{bmatrix} = \begin{bmatrix} 1 & 1 & 0 \\ \frac{1}{\sqrt{2}} & \frac{1}{\sqrt{2}} & 0 \\ 0 & 0 & 1 \end{bmatrix} \begin{bmatrix} \beta_0 \\ \beta_1 \\ \beta_2 \end{bmatrix} + \begin{bmatrix} \epsilon_1 \\ \epsilon_2 \\ \epsilon_3 \end{bmatrix} \qquad (\text{ Model I })$$

and

$$\begin{bmatrix} r_1 \\ r_2 \\ r_3 \end{bmatrix} = \begin{bmatrix} \frac{1}{\sqrt{3}} & \frac{1}{\sqrt{3}} & \frac{1}{\sqrt{3}} \\ \frac{1}{\sqrt{2}} & -\frac{1}{\sqrt{2}} & 0 \\ \frac{1}{\sqrt{6}} & \frac{1}{\sqrt{6}} & -\frac{2}{\sqrt{6}} \end{bmatrix} \begin{bmatrix} \beta_0 \\ \beta_1 \\ \beta_2 \end{bmatrix} + \begin{bmatrix} \epsilon_1 \\ \epsilon_2 \\ \epsilon_3 \end{bmatrix}. \qquad (\text{ Model II })$$

Now if $\sigma^2 = \tau^2 = 1$ since $R'R = I$,

$$C_2 = \begin{bmatrix} \frac{7}{6} & \frac{1}{6} & 0 \\ \frac{1}{6} & \frac{7}{6} & 0 \\ 0 & 0 & 2 \end{bmatrix} \quad \text{and} \quad C_2^+ = \frac{9}{24} \begin{bmatrix} \frac{7}{3} & 0 & 0 \\ -\frac{1}{3} & \frac{7}{3} & 0 \\ 0 & 0 & \frac{4}{3} \end{bmatrix}.$$

For the MSE in (7.2.6)

$$\text{MSE} = p'H \big[A + (\beta - b_2)(\beta - b_2)' \big] Hp,$$

where

$$H = \begin{bmatrix} \frac{63}{72} & 0 & 0 \\ -\frac{1}{8} & \frac{7}{8} & 0 \\ 0 & 0 & \frac{1}{2} \end{bmatrix}$$

and

$$A = \begin{bmatrix} \frac{1}{6} & \frac{1}{6} & 0 \\ \frac{1}{6} & \frac{1}{6} & 0 \\ 0 & 0 & 1 \end{bmatrix}.$$

The exercises below refer to Example 7.2.1.

Exercise 7.2.1. What is the MSE of $\beta_0 + \beta_1 + (1/\sqrt{2})\beta_2$ in terms of β and r? What is the conditional variance?

Exercise 7.2.2. What are the other forms of the MSE neglecting prior information?

7.3 The Comparison of the Conditional MSE of the Bayes Estimator and the Least Square Estimator and the Comparison of the Conditional and the Average MSE

Necessary and sufficient conditions will be derived for a

1. BE to have a smaller MSE than the LS estimator;
2. smaller conditional MSE than average MSE.

It has already been mentioned that the BE has a smaller conditional MSE than the LS estimator for certain values of the β parameters. The results below state that the β parameters must lie within an ellipsoid centered at θ.

In Chapter VI the average MSE was obtained. Without the prior assumptions is the MSE smaller than that averaging over the prior distribution? An ellipsoid will be found inside of which the conditional MSE is less than the average MSE. This ellipsoid is smaller than the one where the BE has a smaller MSE than LS.

To obtain the above mentioned results a region will first be found where the conditional MSE of the BE is less than that of a fixed non-negative matrix N (Theorem 7.3.1). The above mentioned comparisons will then follow as simple corollaries by the proper choice of N. The results may then be further specialized by letting $\theta = 0$ and properly choosing F to compare ridge type estimators.

Theorem 7.3.1. *Let N be any NND matrix with $U'NU$ positive definite. Assume $U'FU$ is PD. Then for the estimable parametric functions $p'\beta$*

$$\text{MSE}(p'\hat{\beta}) \leq p'Np \tag{7.3.1}$$

iff

$$(\beta - \theta)T^{+}(\beta - \theta) \leq 1, \tag{7.3.2}$$

where

$$T = \frac{1}{\sigma^4}UU'F\Gamma_1 N\Gamma_1 FUU' - \frac{1}{\sigma^2}UU'FX'XFUU' \tag{7.3.3}$$

and $U'TU$ is positive definite.

Proof. Equation (7.3.1) holds true iff

$$\text{MSE}(d'U'\hat{\beta}) \leq d'U'NUd \tag{7.3.4}$$

for all d. But

$$U'E(\hat{\beta} - \beta)(\hat{\beta} - \beta)'U$$

$$= (\Lambda + \sigma^2(U'FU)^{-1})^{-1}$$

$$\times [\sigma^2\Lambda + (U'FU)^{-1}U'(\beta - \theta)(\beta - \theta)'U(U'FU)^{-1}\sigma^4]$$

$$\times (\Lambda + \sigma^2(U'FU)^{-1})^{-1}. \tag{7.3.5}$$

Thus (7.3.4) holds true iff

$$U'(\beta - \theta)(\beta - \theta)'U$$

$$\leq U'FU(\Lambda + \sigma^2(U'FU)^{-1})U'NU(\Lambda + \sigma^2(U'FU)^{-1})U'FU\frac{1}{\sigma^4}$$

$$- U'FU\Lambda U'FU\frac{1}{\sigma^2} \tag{7.3.6}$$

$$= U'F(X'X + \sigma^2(UU'FUU')^+)^+N(X'X + \sigma^2(UU'FUU')^+)^+FU\frac{1}{\sigma^4}$$

$$- UU'FX'XFUU'\frac{1}{\sigma^2}. \tag{7.3.7}$$

The result follows from Theorem 2.5.2. ∎

The following corollary compares the conditional MSE of the BE with that of the least square estimator.

Corollary 7.3.1. *The conditional MSE of the BE is less than or equal to that of the LS estimator; i.e.* $\mathrm{MSE}(p'\hat{\beta}) \leq \mathrm{MSE}(p'b)$ *iff*

$$(\beta - \theta)'(2UU'FUU' + \sigma^2(X'X)^+)^+(\beta - \theta) \leq 1. \tag{7.3.8}$$

Proof. Observe that

$$\text{MSE}(p'b) = p'(X'X)^+ p\sigma^2.$$

Let $N = (X'X)^+\sigma^2$. Then $T = 2UU'FUU' + \sigma^2(X'X)^+$ and the result follows. ∎

Example 7.3.1. Comparing BE with LS (Special Case). Consider a linear model where

$$X = \begin{bmatrix} \frac{2}{\sqrt{2}} & \frac{2}{\sqrt{2}} \\ -\frac{1}{\sqrt{2}} & \frac{1}{\sqrt{2}} \end{bmatrix} = I \begin{bmatrix} 2 & 0 \\ 0 & 1 \end{bmatrix} \begin{bmatrix} \frac{1}{\sqrt{2}} & \frac{1}{\sqrt{2}} \\ -\frac{1}{\sqrt{2}} & \frac{1}{\sqrt{2}} \end{bmatrix},$$

$$\theta = \begin{bmatrix} -\frac{1}{2} \\ \frac{1}{2} \end{bmatrix}$$

and

$$F = \begin{bmatrix} \frac{1}{\sqrt{2}} & -\frac{1}{\sqrt{2}} \\ \frac{1}{\sqrt{2}} & \frac{1}{\sqrt{2}} \end{bmatrix} \begin{bmatrix} \frac{\sigma^2}{d_1} & 0 \\ 0 & \frac{\sigma^2}{d_2} \end{bmatrix} \begin{bmatrix} \frac{1}{\sqrt{2}} & \frac{1}{\sqrt{2}} \\ -\frac{1}{\sqrt{2}} & \frac{1}{\sqrt{2}} \end{bmatrix} = UD^{-1}U'.$$

Let $\eta_1 = \beta_1 - 1/2$ and $\eta_2 = \beta_2 + 1/2$. Then (7.3.8) may be written

$$\begin{bmatrix} \eta_1 & \eta_2 \end{bmatrix} U \begin{bmatrix} \frac{\sigma^2}{d_1} + \frac{\sigma^2}{2} & 0 \\ 0 & \frac{\sigma^2}{d_2} + \sigma^2 \end{bmatrix}^{-1} U' \begin{bmatrix} \eta_1 \\ \eta_2 \end{bmatrix} \leq 1. \qquad (7.3.9)$$

Let $\epsilon_1 = (1/\sqrt{2})(\eta_1 - \eta_2)$ and $\epsilon_2 = (1/\sqrt{2})(\eta_1 + \eta_2)$. Then (7.3.9) becomes, after the change of coordinates,

$$\frac{2d_1}{2\sigma^2 + d_1\sigma^2}\epsilon_1^2 + \frac{d_2}{1 + d_2\sigma^2}\epsilon_2^2 \leq 1. \qquad (7.3.10)$$

The length of the axis of the ellipse are

$$\sigma\sqrt{\frac{2+d_1}{2d_1}} \quad \text{and} \quad \sigma\sqrt{\frac{1+d_2}{d_2}}, \quad \text{respectively.}$$

For the contraction estimator $d_1 = 2$, $d_2 = 1$. Then (7.3.10) becomes

$$\frac{4\epsilon_1^2}{3} + \frac{\epsilon_2^2}{2} \leq \sigma^2.$$

Exercise 7.3.1. What is the form of inequality (7.3.10) in Example 7.3.1 for the generalized ridge and contraction estimators and the ordinary ridge estimator?

Exercise 7.3.2. A. What is the center and length of the axis of the ellipsoid in (7.3.8) when

$$X = I \begin{bmatrix} 3 & 0 & 0 \\ 0 & 2 & 1 \\ 0 & 0 & 1 \end{bmatrix} \begin{bmatrix} \frac{1}{\sqrt{3}} & \frac{1}{\sqrt{3}} & \frac{1}{\sqrt{3}} \\ \frac{1}{\sqrt{2}} & -\frac{1}{\sqrt{3}} & 0 \\ \frac{1}{\sqrt{6}} & \frac{1}{\sqrt{6}} & -\frac{2}{\sqrt{6}} \end{bmatrix} = S'\Lambda^{1/2}U',$$

$$\theta' = [1 \; 0 \; -1] \quad \text{and} \quad F = U \begin{bmatrix} d_1 & 0 & 0 \\ 0 & d_2 & 0 \\ 0 & 0 & d_3 \end{bmatrix} U'\sigma^2?$$

B. Answer the question in A above for the special cases of contraction, generalized, contraction, ordinary, and generalized ridge regression estimators.

The conditional MSE of the BE can be compared with the average MSE.

Corollary 7.3.2. *The conditional MSE of the BE is less than or equal to its average MSE; i.e.* $m_2 \leq m_1$ *iff*

$$(\beta - \theta)'(UU'FUU')^{+}(\beta - \theta) \leq 1. \qquad (7.3.11)$$

Proof. Let

$$N = \sigma^2(X'X + \sigma^2(UU'FUU')^{+})^{+}.$$

Then $T = (UU'FUU')^{+}$ and the result follows. ∎

In Section 4.5 it was explained how the various ridge type estimators were special cases of the BE with $\theta = 0$ and F being of the form UDU' for different diagonal matrices D. Theorem 7.3.1 may thus be specialized to give necessary and sufficient conditions for a ridge type estimator to have smaller MSE than LS. It may also be specialized to give necessary and sufficient conditions for the ridge type estimator to have a smaller average than conditional MSE. This leads to the following two additional corollaries to Theorem 7.3.1.

Corollary 7.3.3.
(1) The ordinary ridge regression estimator (4.5.6) has an MSE less than or equal to that of the OLS iff

$$\beta'(\frac{2UU'}{k} + \sigma^2(X'X)^{+})^{+}\beta \leq \sigma^2. \qquad (7.3.12)$$

(2) The generalized ridge regression estimator (4.5.7) has a MSE less than or equal to that of the OLS iff

$$\beta'(2UK^{-1}U' + \sigma^2(X'X)^{+})^{+}\beta \leq \sigma^2. \qquad (7.3.13)$$

(3) The contraction estimator (4.5.8) has a MSE less than or equal to that of the OLS iff

$$\beta'(X'X)\beta \leq \frac{c}{c+1}\sigma^2. \qquad (7.3.14)$$

(4) The generalized contraction estimator (4.5.9) has a MSE less than or equal to that of the OLS iff

$$\beta'(2UC^{-1}\Lambda U' + \sigma^2(X'X)^+)^+\beta \leq \sigma^2. \qquad (7.3.15)$$

Proof. Substitute the form of the prior dispersion that was used to derive the estimators in Section 4.5 into (7.3.8). Recall that for the ordinary ridge estimator $F = (\sigma^2/k)I$, $k > 0$ a positive constant, for the generalized ridge estimator $F = UK^{-1}U'$ (K is a diagonal matrix), for the contraction estimator $F = (\sigma^2/c)(X'X)^+$, $c > 0$ a positive constant, and for the generalized contraction estimator $F = \sigma^2 UC^{-1}\Lambda^{-1}U'$ with C a PD diagonal matrix. ∎

Corollary 7.3.4.

(1) The ordinary ridge regression estimator has conditional MSE less than or equal to average MSE iff

$$\beta'UU'\beta \leq \frac{\sigma^2}{k}. \qquad (7.3.16)$$

(2) The generalized ridge estimator has conditional MSE less than or equal to average MSE iff

$$\beta'UKU'\beta \leq \sigma^2. \qquad (7.3.17)$$

(3) The contraction estimator has conditional MSE less than or equal to average MSE iff

$$\beta'(X'X)\beta \leq \sigma^2. \qquad (7.3.18)$$

(4) The generalized contraction estimator has conditional MSE less than or equal to average MSE iff

$$\beta'UC\Lambda U'\beta \leq \sigma^2. \qquad (7.3.19)$$

Proof. Substitute the different forms of the prior dispersion into (7.3.9). ∎

Farebrother (1976) showed that the ridge estimator of Hoerl and Kennard (1970) has a smaller conditional MSE than the LSE if β lies in the ellipsoid

$$\beta'(\frac{2\sigma^2}{k}I + \sigma^2(X'X)^+)^+\beta < \sigma^2 + \frac{1}{k}\delta'\delta, \tag{7.3.20}$$

where $\delta = V'\beta$.

A more general case of this result is given in Theorem 7.3.2 below.

Example 7.3.2.   Comparison of Conditional with Average MSE. Illustration of Corollary 7.3.4. For the model of Example 7.3.1:

(1) The conditional MSE of the ordinary ridge estimator is less than or equal to the average MSE if

$$\beta_1^2 + \beta_2^2 \leq \frac{\sigma^2}{k}.$$

The parameters lie inside or on the boundary of a circle with radius $\sigma/\sqrt{k}$.

(2) The conditional MSE of the generalized contraction estimator is less than or equal to the average MSE iff

$$k_1\epsilon_1^2 + k_2\epsilon_2^2 \leq \sigma^2.$$

Thus, (ϵ_1, ϵ_2) must lie in the interior or on the boundary of an ellipse with axes $\sigma/\sqrt{k_1}$ and $\sigma/\sqrt{k_2}$.

(3) The conditional MSE of the contraction estimator is less than or equal to the average MSE iff

$$2\epsilon_1^2 + \epsilon_2^2 \leq \sigma^2$$

Thus (ϵ_1, ϵ_2) must lie on the interior or on the boundary of an ellipse with axes $\sigma/\sqrt{2}$ and σ.

(4) The smaller conditional MSE of the generalized contraction estimator is less than or equal to the average MSE iff

$$2c_1\epsilon_1^2 + c_2\epsilon_2^2 \leq \sigma^2.$$

Thus (ϵ_1, ϵ_2) must lie on the interior or on the boundary of the ellipse with axis $\sigma/\sqrt{2c_1}$ and $\sigma/\sqrt{c_2}$.

Exercise 7.3.3. For a design with X as in Example 7.3.2 illustrate Corollary 7.3.4 by giving the geometric properties of the ellipsoids in three-space for each of the four cases as was done in Example 7.3.2 above.

Theorem 7.3.2. Let $F = [U \ V] \begin{bmatrix} A & 0 \\ 0 & B \end{bmatrix} \begin{bmatrix} U' \\ V' \end{bmatrix}$, where A and B are arbitrary PD matrices of appropriate dimensions. Then the conditional MSE of the BE is less than or equal to that of the LS iff

$$(\beta - \theta)'(2F + \sigma^2(X'X)^+)^+(\beta - \theta) \leq \sigma^2 + (\beta - \theta)(VBV')^+(\beta - \theta). \tag{7.3.21}$$

Proof. Notice that $U'FU = A$. Now

$$(\beta - \theta)'(2F + \sigma^2(X'X)^+)^+(\beta - \theta)$$

$$= (\beta - \theta)'[U \ V] \begin{bmatrix} (2A + \sigma^2\Lambda^{-1})^{-1} & 0 \\ 0 & B^{-1} \end{bmatrix} \begin{bmatrix} U' \\ V' \end{bmatrix} (\beta - \theta)$$

$$= (\beta - \theta)'(2UU'FUU' + \sigma^2(X'X)^+)^+(\beta - \theta)$$

$$+ (\beta - \theta)VB^{-1}V'(\beta - \theta)$$

$$< \sigma^2 + (\beta - \theta)VB^{-1}V'(\beta - \theta). \quad \blacksquare \tag{7.3.22}$$

Exercise 7.3.4. Show that when $A = (\sigma^2/k)I$ and $B = I$, (7.3.21) specializes to Farebrother's result (7.3.20).

Above it was shown that the mean square error of $p'\beta$ is less than or equal to $p'Np$, where N is a NND matrix, if the β parameters lie in an ellipsoid. One method of comparison of two estimators would be to consider estimator $p'\hat{\beta}_1$ "better" than $p'\beta_2$ if the ellipsoid where $p'\hat{\beta}_1$ has a smaller MSE than $p'Np$ contains the ellipsoid where $p'\hat{\beta}_2$ has a smaller MSE than $p'Np$. Thus, the following geometric results may be obtained as consequences of Corollaries 7.3.1 and 7.3.2.

Theorem 7.3.3 *If $F_1 \leq F_2$ and $\theta_1 = \theta_2$, then*
  (1) *The ellipsoid inside of which $p'\hat{\beta}_2$ has a smaller MSE than LS contains the ellipsoid where $p'\hat{\beta}_1$ has a smaller MSE than LS.*
  (2) *The ellipsoid where $p'\hat{\beta}_2$ has a smaller conditional than average MSE contains the ellipsoid where $p'\hat{\beta}_1$ has a smaller conditional than average MSE.*
  (3) *If $N_1 \leq N_2$ the ellipsoid where*

$$\mathrm{MSE}(p'\hat{\beta}) \leq p'N_2p$$

contains the ellipsoid where

$$\mathrm{MSE}(p'\hat{\beta}) \leq p'N_1p.$$

Proof. If $A_2 \leq A_1$, then for all β and θ,

$$(\beta - \theta)' A_2 (\beta - \theta) \leq (\beta - \theta)' A_1 (\beta - \theta)$$

by the definition of positive semidefinite matrices. Then clearly if

$$(\beta - \theta)' A_1 (\beta - \theta) \leq 1, \tag{7.3.23}$$

then

$$(\beta - \theta)' A_2 (\beta - \theta) \leq 1. \tag{7.3.24}$$

Thus, from Corollary 7.3.1, since $F_1 \leq F_2$,

$$2U'F_1U + \sigma^2\Lambda^{-1} \leq 2U'F_2U + \sigma^2\Lambda^{-1} \tag{7.3.25}$$

and

$$A_2 = (2UU'F_2UU' + \sigma(X'X)^+)^+ = U(2U'F_2U + \sigma^2\Lambda^{-1})^{-1}U'$$

$$\leq U(2U'F_1U + \sigma^2\Lambda^{-1})^{-1}U' = (2UU'F_1UU' + \sigma^2(X'X)^+)^+$$

$$= A_1. \tag{7.3.26}$$

Thus, 1 is proved. Here, when $F_1 \leq F_2$, $p'\hat{\beta}_2$ is the better estimator.

To establish 2, simply observe that $F_1 \leq F_2$ implies

$$A_2 = (UU'F_2UU')^+ = U(U'F_2U)^{-1}U' \leq U(U'F_1U)^{-1}U' = A_1. \tag{7.3.27}$$

To establish 3, observe that in (7.3.3) $T = USU'$ where

$$S = \frac{1}{\sigma^4}U'F\Gamma_2^+N\Gamma_2^+FU - \frac{1}{\sigma^2}UU'FX'XFUU'. \tag{7.3.28}$$

Let S_i be expression (7.3.28) with N_i in place of N. Now $N_1 \leq N_2$ implies $S_1 \leq S_2$ which in turn implies $T_1 \leq T_2$ and $A_2 = T_2^+ \leq T_1^+ = A_1$. The result then follows. ∎

In Section 4.6.3 the generalized ridge regression estimator was obtained for the case when $U'FU$ was positive semidefinite. It was shown to be a special case of the BE. The inequalities that compare the MSE of the BE and LS and the conditional and average MSE are generalized below.

From Section 4.6 the BE has a smaller MSE than LS iff

$$(\alpha - \eta)'(2\Delta + \sigma^2(Z'Z)^{-1})^{-1}(\alpha - \eta) \leq 1 \tag{7.3.29}$$

or

$$(\beta - \theta)'UP(2\Delta + \sigma^2(Z'Z)^{-1})^{-1}P'U'(\beta - \theta) \leq 1 \tag{7.3.30}$$

or

$$(\beta - \theta)'[2UU'FUU' + \sigma^2UP(Z'Z)^{-1}P'U']^+(\beta - \theta) \leq 1. \tag{7.3.31}$$

Now

$$UP(Z'Z)^{-1}P'U' = UP(P'U'X'XUP)^{-1}P'U'$$

$$= (UPP'U'X'XUPP'U')^{+}$$

$$= (UPP'\Lambda PP'U')^{+}. \qquad (7.3.32)$$

Thus from (7.3.31) and (7.3.32) the MSE of $p'\beta$ is less than that of $p'b$ iff

$$(\beta - \theta)'[2UU'FUU' + \sigma^2(UPP'\Lambda PP'U')^{+}]^{+}(\beta - \theta) \leq 1. \quad (7.3.33)$$

When $U'FU$ is of full rank s, $PP' = I$ and (7.3.30) reduces to (7.3.8).

Likewise the average MSE of $p'\hat{\beta}$ is less than the conditional MSE iff

$$(\alpha - \eta)'\Delta^{-1}(\alpha - \eta) \leq 1 \qquad (7.3.34)$$

or

$$(\beta - \theta)'UP\Delta^{-1}P'U'(\beta - \theta) \leq 1. \qquad (7.3.35)$$

But

$$UP\Delta^{-1}P'U' = U(U'FU)^{+}U' = (UU'FUU')^{+}.$$

Thus (7.3.11) holds even when $U'FU$ is not PD.

The form of the regions where BE have smaller MSE than LS has been given. Now, the analogous results will be obtained for the mixed estimator.

Example 7.3.3. Illustration of Theorem 7.3.3, Statement 1. Consider the linear model in Example 6.5.1 with prior mean $\theta_1 = \theta_2 = 0$ and prior dispersion

$$F_1 = U \begin{bmatrix} 2 & 0 \\ 0 & 1 \end{bmatrix} U'\sigma^2 \quad \text{and} \quad F_2 = U \begin{bmatrix} 4 & 0 \\ 0 & 2 \end{bmatrix} U'\sigma^2.$$

Then $F_2 - F_1 \geq 0$. Let $\gamma = U'\beta$. Thus,

$$\gamma_1 = -\frac{2}{\sqrt{6}}\mu + \frac{1}{\sqrt{6}}\alpha_1 + \frac{1}{\sqrt{6}}\alpha_2,$$

$$\gamma_2 = -\frac{1}{\sqrt{2}}\alpha_1 + \frac{1}{\sqrt{2}}\alpha_2.$$

Then $p'\hat{\beta}_2$ has a smaller MSE than LS if

$$\frac{\gamma_1^2}{14\sigma^2} + \frac{\gamma_2^2}{6\sigma^2} \leq 1.$$

Also $p'\hat{\beta}_1$ has a smaller MSE than LS if

$$\frac{\gamma_1^2}{10\sigma^2} + \frac{\gamma_2^2}{4\sigma^2} \leq 1.$$

Clearly the ellipsoid for $p'\hat{\beta}_2$ has larger axes, thus containing the ellipsoid for $p'\hat{\beta}_1$.

Exercise 7.3.5. Use F_1 and F_2 as given in Example 7.3.3 to illustrate Statement 2 of Theorem 7.3.3.

Example 7.3.4. Illustration of Statement 3 of Theorem 7.3.3. Let X be as in Example 6.5.1. $F = F_1$, $\theta_1 = \theta_2 = 0$. Let

$$N_1 = U \begin{bmatrix} 2 & 0 \\ 0 & 1 \end{bmatrix} U'\sigma^2 \quad \text{and} \quad N_2 = U \begin{bmatrix} 3 & 0 \\ 0 & 2 \end{bmatrix} U'\sigma^2.$$

Now $N_2 - N_1 \geq 0$. Also

$$P_1 = U \begin{bmatrix} \frac{13}{2} & 0 \\ 0 & 3 \end{bmatrix} U'.$$

Now let

$$T_i = \frac{1}{\sigma^4} U U' F_1 P_1 N_i P_1 F_1 U U' - \frac{1}{\sigma^2} U U' F_1 X' X F_1 U U'$$

from (7.3.3). Thus,

$$T_1 = \sigma^2 U \begin{bmatrix} 314 & 0 \\ 0 & 7 \end{bmatrix} U'$$

and

$$T_2 = \sigma^2 U \begin{bmatrix} 483 & 0 \\ 0 & 16 \end{bmatrix} U'.$$

Consequently, $T_2 - T_1 > 0$. The region whose $\text{MSE}(p'\beta_1) \leq N_1$ is

$$\frac{\gamma_1^2}{314\sigma^2} + \frac{\gamma_2^2}{7\sigma^2} \leq 1$$

and where $\text{MSE}(p'\beta_2) \leq N_2$ is

$$\frac{\gamma_1^2}{483\sigma^2} + \frac{\gamma_2^2}{13\sigma^2} \leq 1.$$

Clearly the second region contains the first.

Exercise 7.3.6. Let X be as in Exercise 7.3.2,

$$F = U \begin{bmatrix} 3 & 0 & 0 \\ 0 & 2 & 0 \\ 0 & 0 & 1 \end{bmatrix} U', \quad N_1 = U \begin{bmatrix} 5 & 0 & 0 \\ 0 & 4 & 0 \\ 0 & 0 & 0 \end{bmatrix} U',$$

and

$$N_2 = U \begin{bmatrix} 3 & 0 & 0 \\ 0 & 1 & 0 \\ 0 & 0 & 0 \end{bmatrix} U'.$$

Illustrate Statement 3 of Theorem 7.3.3 as was done in Example 7.3.3 above.

Exercise 7.3.7. Let X be such that $X'X = \begin{bmatrix} 3 & 0 & 0 \\ 0 & 2 & 0 \\ 0 & 0 & 1 \end{bmatrix}$. For each ridge type estimator obtain the length of the axis of the ellipsoid where the estimator has smaller MSE than LS. What is the equation of the ellipsoid?

Exercise 7.3.8. Show that for any ellipsoid

$$\beta' A \beta \le 1$$

there is a transformation of coordinates $\gamma = p'\beta$ so that the ellipsoid takes the form

$$\sum_{i=1}^{n} \frac{\gamma_i^2}{a_i} \le 1.$$

Interpret the result in the context of comparing ridge type and LS estimators.

Example 7.3.5. Comparison of BE with LS. Let X, F, and θ be as in Example 4.6.1. Now

$$\alpha_1 = \frac{3}{\sqrt{12}}\mu + \frac{1}{\sqrt{12}}(\tau_1 + \tau_2 + \tau_3),$$

$$\alpha_2 = -\frac{2}{\sqrt{6}}\tau_1 + \frac{1}{\sqrt{6}}(\tau_2 + \tau_3),$$

$$\eta_1 = \frac{6}{\sqrt{12}}, \quad \eta_2 = 0,$$

$$Z'Z = \begin{bmatrix} 12 & 0 \\ 0 & 3 \end{bmatrix} \quad \text{and} \quad \Delta = \begin{bmatrix} 1 & 0 \\ 0 & 1 \end{bmatrix}.$$

Thus, the BE has a smaller MSE than LS if

$$\begin{bmatrix} \alpha_1 - \frac{6}{\sqrt{12}}, & \alpha_2 \end{bmatrix} \left[\begin{bmatrix} 2 & 0 \\ 0 & 2 \end{bmatrix} + \sigma^2 \begin{bmatrix} \frac{1}{12} & 0 \\ 0 & \frac{1}{3} \end{bmatrix} \right]^{-1} \begin{bmatrix} \alpha_1 - \frac{6}{\sqrt{12}} \\ \alpha_2 \end{bmatrix} \leq 1.$$

Suppose $\sigma^2 = 12$. Then the ellipsoid of LS optimality would be

$$\frac{\left(\alpha_1 - (6/\sqrt{12}) \right)^2}{3} + \frac{\alpha_2^2}{6} \leq 1.$$

The MSE is less than the average MSE when

$$\left(\alpha_1 - \frac{6}{\sqrt{12}} \right)^2 + \alpha_2^2 \leq 1.$$

Exercise 7.3.9. Reformulate above results in terms of μ and τ_i.

7.4 The Comparison of the MSE of a Mixed Estimator with the LS Estimators

In Section 4.4 it was observed that the augmented linear model could be written as two separate linear models and that the LS estimator could be found with respect to each one. These MSEs were

$$\text{MSE}(p'b_1) = p'(X'X)^+p\sigma^2 \qquad (7.4.1)$$

and

$$\text{MSE}(p'b_2) = p'(R'R)^+p\tau^2. \qquad (7.4.2)$$

Now the MSE of the mixed estimators and the least square estimators will be compared neglecting prior assumptions. The following analogue of Theorem 7.3.1 compares the MSE of the BE with an NND matrix. Choosing the matrix correctly then gives comparisons with (7.4.1) and (7.4.2) as special cases.

Theorem 7.4.1. *Consider Model I as defined in Section 4.4. Assume $R = AU'$ where A is a PD matrix. Let N be any NND matrix. Then*

$$\text{MSE}(p'\hat{\beta}) \leq p'Np \qquad (7.4.3)$$

iff

$$(R\beta - r)'T^{-1}(R\beta - r) \leq \frac{1}{\sigma^4}, \qquad (7.4.4)$$

where

$$T = R'^+ M R^+ \qquad (7.4.5)$$

and

$$M = \Gamma_2 N \Gamma_2 - \sigma^2 \tau^4 X'X. \qquad (7.4.6)$$

Proof. From (7.2.3) for estimable parametric functions
$\mathrm{MSE}(p'\hat{\beta})$

$$= d' \, (\Lambda\tau^2 + A'A\sigma^2)^{-1} [\tau^4\sigma^2\Lambda + \sigma^4 A'(AU'\beta - r)(AU'\beta - r)'A]$$

$$\times \, (\Lambda\tau^2 + A'A\sigma^2)^{-1} d. \tag{7.4.7}$$

Thus (7.4.3) holds true if

$$(AU'\beta - r)'AS^{-1}A'(AU'\beta - r) \leq \frac{1}{\sigma^4}, \tag{7.4.8}$$

where $S = [(\tau^2\Lambda + \sigma^2 A'A)U'NU(\tau^2\Lambda + \sigma^2 A'A) - \sigma^2\tau^4\Lambda)]1/\sigma^4$.
Now

$$\begin{aligned} S &= U'USU'U \\ &= U'(\tau^2 X'X + \sigma^2 R'R)N(\tau^2 X'X + \sigma^2 R'R) - U'X'XU\sigma^2 \\ &= U'MU, \end{aligned} \tag{7.4.9}$$

where

$$M = [(\tau^2 X'X + \sigma^2 R'R)N(\tau^2 X'X + \sigma^2 R'R) - \sigma^2\tau^4 X'X]\frac{1}{\sigma^4}. \tag{7.4.10}$$

Also

$$AS^{-1}A' = [A'^{-1}SA^{-1}]^{-1} = [A'^{-1}U'MUA^{-1}]^{-1} = [R'^{+}MR^{+}]^{-1}. \tag{7.4.11}$$

To compare the estimate with b_1 let $N = \sigma^2(X'X)^{+}$. Then

$$M = [2\sigma^4\tau^2 RR' + \sigma^6 R'R(X'X)^{+}R'R]\frac{1}{\sigma^4} \tag{7.4.12}$$

$$= \sigma^2 R'R(X'X)^{+}R'R + 2\tau^4 RR'$$

and

$$T = \sigma^2 R(X'X)^{+}R' + 2\tau^2 I. \quad \blacksquare \tag{7.4.13}$$

Thus, from Theorem 7.4.1:

Corollary 7.4.1. *The*

$$\text{MSE}(p'\hat{\beta}) \le \text{MSE}(p'b_1)$$

iff

$$(R\beta - r)'(\sigma^2 R(X'X)^+ R' + 2\tau^2 I)^{-1}(R\beta - r) \le 1. \qquad (7.4.14)$$

To compare the MSE of $p'\hat{\beta}$ with $p'b_2$ neglecting prior information let

$$N = (\tau^2(X'X) + \sigma^2(R'R))^+ \sigma^2 \tau^2. \qquad (7.4.15)$$

Then $M = \tau^2(R'R)$ and $T = \tau^2 UU'$. Thus:

Corollary 7.4.2. *The*

$$\text{MSE}(p'\hat{\beta}) \le \text{MSE}(p'b_2) \qquad (7.4.16)$$

iff

$$(R\beta - r)'UU'(R\beta - r) \le \tau^2. \qquad (7.4.17)$$

Now consider Model II. Exchanging of sample and prior information enables the statement of the Theorem 7.4.2.

Example 7.4.1. Illustration of Theorem 7.4.1 and its Corollaries. Consider the design of Example 6.5.1.

$$\text{Let } R = \begin{bmatrix} \sqrt{6} & 0 \\ 0 & \sqrt{2} \end{bmatrix} \begin{bmatrix} -\frac{2}{\sqrt{6}} & \frac{1}{\sqrt{6}} & \frac{1}{\sqrt{6}} \\ 0 & -\frac{1}{\sqrt{2}} & \frac{1}{\sqrt{2}} \end{bmatrix} = \begin{bmatrix} 2 & 1 & 1 \\ 0 & -1 & 1 \end{bmatrix}.$$

$N = I$, $\sigma^2 = \tau^2 = 1$. Then

$$P_2 = \begin{bmatrix} 4 & 2 & 2 \\ 2 & 2 & 0 \\ 2 & 0 & 2 \end{bmatrix} + \begin{bmatrix} 4 & -2 & -2 \\ -2 & 2 & 0 \\ -2 & 0 & 2 \end{bmatrix} = \begin{bmatrix} 8 & 0 & 0 \\ 0 & 4 & 0 \\ 0 & 0 & 4 \end{bmatrix}$$

and from Equation (7.4.6)

$$M = \begin{bmatrix} 64 & 0 & 0 \\ 0 & 16 & 0 \\ 0 & 0 & 16 \end{bmatrix} - \begin{bmatrix} 4 & 2 & 2 \\ 2 & 2 & 0 \\ 2 & 0 & 2 \end{bmatrix} = \begin{bmatrix} 60 & -2 & -2 \\ -2 & 14 & 0 \\ -2 & 0 & 14 \end{bmatrix}.$$

Now

$$R^+ = \begin{bmatrix} \frac{2}{\sqrt{6}} & 0 \\ \frac{1}{\sqrt{6}} & -\frac{1}{\sqrt{2}} \\ \frac{1}{\sqrt{6}} & \frac{1}{\sqrt{2}} \end{bmatrix} \begin{bmatrix} \frac{1}{\sqrt{6}} & 0 \\ 0 & \frac{1}{\sqrt{2}} \end{bmatrix} = \begin{bmatrix} \frac{1}{3} & 0 \\ \frac{1}{6} & -\frac{1}{2} \\ \frac{1}{6} & \frac{1}{2} \end{bmatrix}.$$

Then

$$T = \begin{bmatrix} \frac{1}{3} & \frac{1}{6} & \frac{1}{6} \\ 0 & -\frac{1}{2} & \frac{1}{2} \end{bmatrix} \begin{bmatrix} 60 & -2 & -2 \\ -2 & 14 & 0 \\ -2 & 0 & 14 \end{bmatrix} \begin{bmatrix} \frac{1}{3} & 0 \\ \frac{1}{6} & -\frac{1}{2} \\ \frac{1}{6} & \frac{1}{2} \end{bmatrix} = \begin{bmatrix} \frac{65}{9} & 0 \\ 0 & 7 \end{bmatrix}.$$

The condition in (7.4.4) is

$$\frac{65}{9}(2\mu + \alpha_1 + \alpha_2 - r_1)^2 + \frac{1}{7}(\alpha_2 - \alpha_1 - r_2)^2 \le 1;$$

i.e., the estimable parametric functions lie on an ellipse centered at (r_1, r_2) with axis $\sqrt{9/65}$ and $\sqrt{7}$ when $\mathrm{MSE}(p'\hat{\beta}) \le \mathrm{MSE}(p'\beta)$.
Since

$$R(X'X)^+ R' = A\Lambda^{-1} A = \begin{bmatrix} \sqrt{6} & 0 \\ 0 & \sqrt{2} \end{bmatrix} \begin{bmatrix} \frac{1}{6} & 0 \\ 0 & \frac{1}{2} \end{bmatrix} \begin{bmatrix} \sqrt{6} & 0 \\ 0 & \sqrt{2} \end{bmatrix} = I,$$

the mixed estimator has a smaller MSE than LS iff

$$(2\mu + \alpha_1 + \alpha_2 - r_1)^2 + (\alpha_2 - \alpha_1 - r_1)^2 \leq 3.$$

Thus, the estimable parametric functions lie on a circle whose radius is $\sqrt{3}$.

Theorem 7.4.2. *Consider Model II with $P'X'XP$ positive definite. Let N be any NND matrix. Then*

$$\text{MSE}(p'\hat{\beta}) \leq p'Np \qquad (7.4.18)$$

iff

$$(X\beta - Y)'T^{-1}(X\beta - Y) \leq \frac{1}{\tau^4}, \qquad (7.4.19)$$

where

$$T = X'^+ M X^+ \qquad (7.4.20)$$

and

$$M = \Gamma_2 N \Gamma_2 - \tau^4 R' R. \qquad (7.4.21)$$

Likewise:

Corollary 7.4.3. *The*

$$\text{MSE}(p'\hat{\beta}) \leq \text{MSE}(p'b_2) \qquad (7.4.22)$$

iff

$$(X\beta - Y)'(\tau^2 X(R'R)^+ X' + 2\sigma^4 PP')^+ (X\beta - Y) \leq 1. \quad (7.4.23)$$

Also:

Corollary 7.4.4. *The*

$$\text{MSE}(p'\hat{\beta}) \leq \text{MSE}(p'b_1) \quad (7.4.24)$$

iff

$$(X\beta - Y)'PP'(X\beta - Y) \leq \sigma^2. \quad (7.4.25)$$

Thus, the nature of the comparisons of the conditional MSE are similar for the BE and the mixed estimator.

Exercise 7.4.1. Use X and R in Example 7.4.1 above to calculate the relevant quantities in Theorem 7.4.2 and its corollary.

Exercise 7.4.2. Find necessary and sufficient conditions for the weighted least square estimator derived in Exercises 3.3.7 and 3.3.9 to have a smaller MS than ordinary LS. How does the MSE depend on w? Which weight w if any is optimal?

7.5 The Comparison of the MSE of Two BE

In Section 7.1 it was observed that

$$m_2 = \text{MSE}(p'\beta) = \text{Variance}(p'\hat{\beta}) + [\text{Bias}(p'\hat{\beta})]^2. \quad (7.5.1)$$

Let $p'\hat{\beta}_1$ and $p'\hat{\beta}_2$ be two BE of $p'\beta$. Let

$$C(p'\hat{\beta}_1, \ p'\hat{\beta}_2) = \text{Variance}(p'\hat{\beta}_2) - \text{Variance}(p'\hat{\beta}_1). \quad (7.5.2)$$

The following theorem of Trenkler (1985) gives necessary and sufficient conditions for $p'\hat{\beta}_1$ to have a MSE less than or equal to that of $p'\hat{\beta}_2$. It is a direct consequence of Theorem 2.5.3.

Theorem 7.5.1. *The*

$$\text{MSE}(p'\hat{\beta}_1) \leq \text{MSE}(p'\hat{\beta}_2) \tag{7.5.3}$$

iff

(1) $C(\hat{\beta}_1, \hat{\beta}_2) + \text{Bias}(\hat{\beta}_1) \text{Bias}(\hat{\beta}_2)'$ *is NND.*
(2) The bias of $\hat{\beta}_1$ *is in the range of*

$$T = C(\hat{\beta}_1, \hat{\beta}_2) + \text{Bias}(\hat{\beta}_2) \text{Bias}(\hat{\beta}_2)'$$

and

$$(\text{Bias } \hat{\beta}_1)'T^+(\text{Bias}\hat{\beta}_1) \leq 1. \tag{7.5.4}$$

When $\hat{\beta}_2 = b_2$, *i.e.,* $\hat{\beta}_2$ *is the LS estimator, inequality (7.5.4) reduces to (7.3.8).*

Notice that T depends of β as well as F_1 and F_2. When the BE was compared with LS the matrix T did not depend on the β parameters. Thus, the shape of the region where $p'\hat{\beta}_1$ has a smaller MSE than $p'\hat{\beta}_2$ is generally more complicated than an ellipsoid.

Exercise 7.5.1. Show that the ellipsoid in (7.5.4) reduces to that of (7.3.8) when $\hat{\beta}_2 = b$.

Exercise 7.5.2. Consider two contraction estimators $\hat{\beta}_{c_1}$ and $\hat{\beta}_{c_2}$ with parameters k_1 and k_2. Assume $k_1 > k_2$. Is there an ellipsoid where $\text{MSE}(p'\hat{\beta}_{c_1}) \leq \text{MSE}(p'\hat{\beta}_{c_2})$? If so find it.

However in certain cases easier comparisons can be made of the variances of $p'\beta_1$ and $p'\beta_2$, the MSE averaging over certain quadratic loss functions of interest and the ratio of the weighted squared biases. First a comparison of the variances will be made.

Conditions for $C(p'\hat{\beta}_1, p'\hat{\beta}_2) \geq 0$, where $p'\hat{\beta}_1$ is the BE with prior dispersion F_1 and $p'\hat{\beta}_2$ is the BE with prior dispersion F_2 will now be obtained.

Theorem 7.5.2. *Suppose* $F_1 = [U \quad V] \begin{bmatrix} D_2 & C_2 \\ C_2' & E_2 \end{bmatrix} \begin{bmatrix} U' \\ V \end{bmatrix}$ *and* $F_2 = [U \quad V] \begin{bmatrix} D_2 & C_2 \\ C_2' & E_2 \end{bmatrix} \begin{bmatrix} U' \\ V' \end{bmatrix}$. *Also* D_1 *and* D_2 *are diagonal matrices,* C_1, C_2, E_1 *and* E_2 *are arbitrary matrices. Also* $F_2 - F_2 \geq 0$. *Assume* $p'\beta$ *is estimable. Then* $C(p'\hat{\beta}_1, p'\hat{\beta}_2) \geq 0$.

Proof. From (7.1.1) for $i = 1,2$

$$\text{Var}(p'\hat{\beta}_j) = d' R_j d, \tag{7.5.5}$$

where

$$R_j = [\Lambda + \sigma^2 D_j^{-1}]^{-1} \Lambda [\Lambda + \sigma^2 D_j^{-1}]^{-1}.$$

It will suffice to show $R_2 - R_1 \geq 0$ if $D_2 - D_1 \geq 0$. Since the matrix D_j is diagonal with diagonal elements d_{ij}, R_j is diagonal with elements

$$f(d_{ij}) = \left[\frac{1}{\lambda_i + \frac{\sigma^2}{d_{ij}}} \right]^2 \lambda_i = \frac{d_{ij}^2 \lambda_i}{(\lambda_i d_{ij} + \sigma^2)^2}. \tag{7.5.6}$$

Now the derivative

$$f'(d_{ij}) = \frac{2 d_{ij} \lambda_i \sigma^2}{(\lambda_i d_{ij} + \sigma^2)^3} > 0. \tag{7.5.7}$$

Since d_{ij} and λ_i are positive numbers $f(d_{ij})$ is an increasing function. Thus if $D_2 - D_1 \geq 0$ then $R_2 - R_1 \geq 0$ and

$$C(p'\beta_1, p'\beta_2) = p'(R_2 - R_1)p \geq 0.$$

Example 7.5.1. Importance of Hypothesis of Theorem 7.5.2.
The counterexample in Example 6.6.2 can be used to show that the conclusion of Theorem 7.5.2 may not hold if F_1 and F_2 are not in the given form. Let

$$X'X = I, \quad \sigma^2 = 1, \quad F_1 = \begin{bmatrix} 2 & 1 \\ 1 & 1 \end{bmatrix}, \quad F_2 = \begin{bmatrix} 2 & 1 \\ 1 & 3 \end{bmatrix}.$$

From (7.1.1) it suffices to show that $(I + F_2^{-1})^{-2} - (I + F_1^{-1})^{-2}$ is not NND.

Now since

$$(I + F_1^{-1})^{-2} = \frac{1}{25} \begin{bmatrix} 10 & 5 \\ 5 & 5 \end{bmatrix}$$

and

$$(I + F_2^{-1})^{-2} = \frac{5}{121} \begin{bmatrix} 10 & 3 \\ 3 & 13 \end{bmatrix}$$

then

$$J = (I + F_2^{-1})^{-2} - (I + F_1^{-1})^{-2} = \frac{1}{605} \begin{bmatrix} 8 & -46 \\ -46 & 204 \end{bmatrix}.$$

Since the determinant of J is negative it is not a positive semidefinite matrix.

Exercise 7.5.3. Let $A = \begin{bmatrix} 2 & 1 \\ 1 & 1 \end{bmatrix}$ and $B = \begin{bmatrix} 2 & 1 \\ 1 & 1 + \epsilon \end{bmatrix}$ with $\epsilon > 0$.
Show that $B - A$ is positive semidefinite but $B^2 - A^2$ is not positive semidefinite.

Now comparison of the conditional MSE averaging over a quadratic loss function will be done. Special attention will be paid to the quadratic loss functions important in the statistical literature. These include the individual MSE, total MSE and the MSE averaging over a predictive loss function.

Let A be a NND matrix. If V is the conditional dispersion matrix of $\hat{\beta}$ then the weighted variance of $\hat{\beta}$ is defined to be

$$\text{tr } AV = \text{tr } AE(\hat{\beta} - E(\hat{\beta}))(\hat{\beta} - E(\hat{\beta}))'$$

$$= E(\hat{\beta} - E(\hat{\beta}))'A(\hat{\beta} - E(\hat{\beta})).$$

(7.5.8)

From Theobald's result it is easily seen that if $F_2 - F_1 > 0$, where F_1 and F_2 are in the form of Theorem 7.5.2,

$$\text{tr } AV_1 \leq \text{tr } AV_2. \qquad (7.5.9)$$

The weighted squared bias is

$$\text{tr } AB = \text{tr } A(E(\hat{\beta}) - \beta)(E(\hat{\beta}) - \beta)'$$

$$= [E(\hat{\beta}) - \beta]'A[E(\hat{\beta}) - \beta].$$

(7.5.10)

An estimator $p'\hat{\beta}_1$ is more efficient than $p'\hat{\beta}_2$ if it has a larger signal to noise ratio, i.e.

$$\frac{p'B_1 p}{p'V_1 p} \geq \frac{p'B_2 p}{p'V_2 p}. \qquad (7.5.11)$$

However, it is impossible to compare the MSE of $\hat{\beta}_1$ and $\hat{\beta}_2$ by virtue of Theorem 2.5.3 of Teräsvirta. Nevertheless, it is possible to make comparisons for the weighted signal to noise ratio $(\text{tr } AB)/(\text{tr } AV)$ for certain choices of A.

Theorem 7.5.3. *Suppose F_1 and F_2 are in the form of Theorem 7.5.2. Let $A = U\Delta U' + VV'$ where Δ is a diagonal matrix. Let B_1 and B_2 be the squared matrix bias of $\hat{\beta}_1$ and $\hat{\beta}_2$ respectively. If $F_1 \leq F_2$ then $\text{tr } B_1 A \leq \text{tr } B_2 A$.*

Proof. The matrix

$$H = (X'X + \sigma^2(UU'FUU')^+)^+(UU'FUU')^+\sigma^2 + VV'$$
$$(7.5.12)$$
$$= U(\Lambda + \sigma^2 D^{-1})^{-1}D^{-1}U'\sigma^2 + VV'.$$

The individual diagonal elements of $(\Lambda + \sigma^2 D^{-1})^{-1}D^{-1}$ are

$$\theta_j = (\lambda_j + \frac{\sigma^2}{d_j})^{-1}\frac{1}{d_j} = \frac{1}{\lambda_j d_j + \sigma^2}. \qquad (7.5.13)$$

Let H_i be (7.5.12) with F_i and D_i in place of F and D respectively. If D_i has diagonal elements d_{ij}, since $F_1 \leq F_2$ iff $d_{1j} \leq d_{2j}$ for all j, $H_1 \leq H_2$. Now this implies,

$$\text{tr } B_1 A = (\beta - \theta)'H_1' A H_1(\beta - \theta) \leq (\beta - \theta)'H_2' A H_2(\beta - \theta) = \text{tr } B_2 A,$$
$$(7.5.14)$$

because

$$H_1 A H_1 = U\theta_1 \Delta\theta_1 U' + VV' \leq U\theta_2 \Delta\theta_2 U' + VV' = H_2 A H_2. \quad \blacksquare$$
$$(7.5.15)$$

From Theorem 7.5.3 and (7.5.8) it follows that when F_i are as in Theorem 7.5.2

$$\frac{\text{tr } AB_1}{\text{tr } AV_1} \geq \frac{\text{tr } AB_2}{\text{tr } AV_2}. \qquad (7.5.16)$$

Thus more precise prior information means:

1. a smaller weighted conditional variance;
2. a larger weighted bias;
3. a larger signal to noise ratio.

For an NND matrix A where $\hat\theta$ is an m dimensional random variable estimating the m dimensional vector θ the MSE (2.4.2) is the sum of the weighted variance and the weighted bias. Thus,

$$M_A = E[\hat\theta - E(\hat\theta))'A(\hat\theta - E(\hat\theta)) + (\theta - E(\hat\theta))'A(\theta - E(\hat\theta)).$$
$$(7.5.17)$$

The important cases of the

$$\text{MSE} = E(\hat{\beta} - \beta)'A(\hat{\beta} - \beta) \qquad (7.5.18a)$$

that are frequently treated in the literature are those with $A = X'X$ (the predictive MSE), $A = I$, the total MSE, and the individual MSE of the components of $U'\beta$. When $A = X'X$,

$$\text{MSE} = E(\hat{\beta} - \beta)'X'X(\hat{\beta} - \beta). \qquad (7.5.18b)$$

In a linear regression model the predicted values are estimated by $\hat{Y} = X\hat{\beta}$, hence the name predictive loss function. When $A = I$

$$\text{MSE} = E(\hat{\beta} - \beta)'(\hat{\beta} - \beta), \qquad (7.5.18c)$$

the total MSE. When $A = U_j U_j'$, where U_j is the jth column of U,

$$\text{MSE} = E(\hat{\beta} - \beta)'U_j U_j'(\hat{\beta} - \beta)$$

$$\qquad\qquad\qquad\qquad (7.5.18d)$$

$$= E(\hat{\gamma}_j - \gamma_j)^2,$$

where $\gamma_j = U_j'\beta$.

Exercise 7.5.4. Let $a = \begin{bmatrix} 3 \\ -2 \\ 4 \end{bmatrix}$, $b = \begin{bmatrix} -2 \\ 1 \\ 3 \end{bmatrix}$.

A. Find $M = aa' - bb'$.
B. Find $\text{tr}(aa' - bb')$. Notice that your answer is positive.
C. Find the eigenvalues of M. Observe that the positive eigenvalue has a larger absolute value.

Exercise 7.5.5. Given n dimensional column vectors b and c. Show that

A. $\text{tr}(bb' - cc') = (b'b - c'c) = \text{tr } bb' - \text{tr } cc'$.
B. $\det(bb' - cc') = (b'c)^2 - b'bc'c \leq 0$.

Exercise 7.5.6. Let $p'\hat{\beta}_1$ and $p'\hat{\beta}_2$ be BE derived from two priors with dispersions F_1 and F_2 that satisfy $F_1 \leq F_2$. Let the SVD of

$$M = B_1 - B_2 = [P_1 \ P_2] \begin{bmatrix} \lambda_+ & 0 \\ 0 & \lambda_- \end{bmatrix} \begin{bmatrix} P_1' \\ P_2' \end{bmatrix},$$

where λ_+ and λ_- are the positive and the negative eigenvalues of $B_1 - B_2$. Show that

A. $p'Mp = |\lambda_+|p'P_1P_1'p - |\lambda_-|p'P_2P_2'p$, where $|\lambda_+| > |\lambda_-|$.
B. If $p'\beta$ is a parametric function where

$$\frac{p'P_2P_2'p}{P'P_1P_1'p} \leq \frac{|\lambda_+|}{|\lambda_-|}$$

then $p'Mp \geq 0$. Otherwise $p'Mp < 0$.

Exercise 7.5.7. Let

$$b = \begin{bmatrix} 1 \\ 2 \end{bmatrix}, \qquad c = \begin{bmatrix} 3 \\ 4 \end{bmatrix}, \qquad d = \begin{bmatrix} 4 \\ 5 \end{bmatrix}.$$

Show that the difference between the eigenvalues of $d'd - b'b$ is greater than that between $c'c - b'b$.

Exercise 7.5.8. Let $p'\beta_i$ be derived from priors with mean zero and dispersion UD_iU' with D_i diagonal matrices. Let $F_1 \leq F_2 \leq F_3 \leq F_4$. Show that for $B_4 - B_1$ the difference in the absolute values of the two eigenvalues is greater than that of $B_3 - B_2$.

A necessary and sufficient condition is given in Theorem 7.5.4 below for one BE to have a smaller predictive or total MSE than another.

Theorem 7.5.4. *Suppose $F_1 \leq F_2$, where F_i is of the form of Theorem 7.5.2. Then if $A = UDU' + VV'$ with D a NND diagonal* matrix the MSE (7.5.18) of $\hat{\beta}_1$ is less than that of $\hat{\beta}_2$ iff $\gamma_j = U_j'\beta$ lies in the s dimensional rectangle

$$|\gamma_j - U_j'D| \le \left[\frac{(d_{1j} + d_{2j})\sigma^2 + 2\lambda_j d_{1j}d_{2j}}{2\sigma^2 + \lambda_j(d_{1j} + d_{2j})}\right]^{1/2}, \quad 1 \le j \le s. \quad (7.5.19)$$

When the jth inequality holds true the MSE of $U_j'\beta_1$ is less than that of $U_j'\beta_2$.

Proof. Let $\hat{\gamma} = U'\hat{\beta}$. Then

$$E_\beta(\hat{\beta} - \beta)'A(\hat{\beta} - \beta) = E_\beta(\hat{\gamma} - \gamma)'D(\hat{\gamma} - \gamma) + E_\beta(\hat{\beta} - \beta)'VV'(\hat{\beta} - \beta). \tag{7.5.20}$$

Since $\hat{\beta} = \theta + FX'(XFX' + \sigma^2 I)^{-1}(Y - X\theta)$ when $F = UDU'$, $V'\hat{\beta} = V'\theta$. Thus,

$$E_\beta(\hat{\beta} - \beta)'VV'(\hat{\beta} - \beta) = (\theta - \beta)'VV'(\theta - \beta). \tag{7.5.21}$$

It suffices to show the individual MSE of the components of $\hat{\gamma}_1$ is less than that of those of $\hat{\gamma}_2$ by Theorem 2.4.2. Let U_j be the columns of U. Now, if $i = 1, 2$,

$$E(\hat{\gamma}_{ij} - \gamma_{ij})^2 = U_j'E(\hat{\beta} - \beta)(\hat{\beta} - \beta)U_j. \tag{7.5.22}$$

From (7.1.6) for $i = 1, 2$,

$$E(\hat{\gamma}_{ij} - \gamma_{ij})^2 = \left(\frac{1}{\lambda_j + \frac{\sigma^2}{d_{ij}}}\right)\sigma^2\lambda_j + \frac{\sigma^4}{d_{ij}^2}(\gamma_j - U_j'\theta)^2\left(\frac{1}{\lambda_j + \frac{\sigma^2}{d_{ij}}}\right)$$

$$= \left(\frac{1}{\lambda_j d_{ij} + \sigma^2}\right)^2 (\sigma^2\lambda_j d_{ij}^2 + \sigma^4(\gamma_j - U_j'\theta)^2). \tag{7.5.23}$$

If $F_1 \le F_2$ then $d_{1j} \le d_{2j}$. Also

$$\left(\frac{d_{2j}}{\lambda_j d_{2j} + \sigma^2}\right)^2 - \left(\frac{d_{1j}}{\lambda_j d_{1j} + \sigma^2}\right)^2$$

$$= \left(\frac{d_{2j}}{\lambda_j d_{2j} + \sigma^2} + \frac{d_{1j}}{\lambda_j d_{1j} + \sigma^2} \right) \left(\frac{d_{2j}}{\lambda_j d_{2j} + \sigma^2} - \frac{d_{1j}}{\lambda_j d_{1j} + \sigma^2} \right)^2$$

$$= \frac{[2\lambda_j d_{1j} d_{2j} + \sigma^2 (d_{1j} + d_{2j})](d_{2j} - d_{1j})\sigma^2}{(\lambda_j d_{2j} + \sigma^2)^2 (\lambda_j d_{1j} + \sigma^2)^2}. \tag{7.5.24}$$

Also

$$\left(\frac{1}{\lambda_j d_{1j} + \sigma^2} \right)^2 - \left(\frac{1}{\lambda_j d_{2j} + \sigma^2} \right)^2$$

$$= \left(\frac{1}{\lambda_j d_{1j} + \sigma^2} + \frac{1}{\lambda_j d_{2j} + \sigma^2} \right) \left(\frac{1}{\lambda_j d_{1j} + \sigma^2} - \frac{1}{\lambda_j d_{2j} + \sigma^2} \right)^2$$

$$= \frac{\lambda_j (d_{2j} - d_{1j})[\lambda(d_{1j} + d_{2j}) + 2\sigma^2]}{(\lambda_j d_{1j} + \sigma^2)^2 (\lambda_j d_{2j} + \sigma^2)^2}. \tag{7.5.25}$$

Thus

$$E(\hat{\gamma}_{1j} - \gamma_{1j})^2 \leq E(\hat{\gamma}_{2j} - \gamma_{2j})^2 \tag{7.5.26}$$

iff

$$\sigma^4 \lambda_j (d_{2j} - d_{1j})[\lambda_j (d_{1j} + d_{2j}) + 2\sigma^2](\gamma_j - U_j'\theta)^2$$

$$\tag{7.5.27}$$

$$\leq \sigma^2 \lambda_j [2\lambda_j d_{1j} d_{2j} + \sigma^2 (d_{1j} + d_{2j})](d_{2j} - d_{1j})\sigma^2.$$

The result follows if $d_{1j} \neq d_{2j}$. ∎

Example 7.5.2. Comparison of Individual MSE of BE. Let X, F_1 and F_2 be as in Example 7.3.3. Let $\sigma^2 = 1$. Then for (7.5.9)

$$|\gamma_1| \leq 1.63,$$
$$|\gamma_2| \leq 1.48$$

or
$$-1.63 \leq \gamma_1 \leq 1.63,$$
$$-1.48 \leq \gamma_2 \leq 1.48$$

is the rectangle inside of which $\hat{\beta}_1$ has smaller individual total and predictive MSE than $\hat{\beta}_2$.

The result of Theorem 7.5.1 is a bit cumbersome. A simpler necessary but not sufficient condition for $p'\hat{\beta}_1$ to have a smaller MSE than $p'\hat{\beta}_2$ is:

Theorem 7.5.5. *Let F_1 and F_2 be in the form of Theorem 7.5.1 with $F_1 \leq F_2$. Let $p'\hat{\beta}_1$ and $p'\hat{\beta}_2$ be the BE with prior dispersion F_1 and F_2. Suppose β lies in the ellipsoid*

$$(\text{Bias } \hat{\beta}_1)'[C(\hat{\beta}_1, \hat{\beta}_2)]^+(\text{Bias } \hat{\beta}_1) \leq 1; \qquad (7.5.28)$$

then

$$\text{MSE } (p'\hat{\beta}_1) \leq \text{MSE } (p'\hat{\beta}_2). \qquad (7.5.29)$$

Proof. Note that the ellipsoid in (7.5.28) belongs to the one in (7.5.4).∎

Exercise 7.5.9. Let

$$X'X = \begin{bmatrix} 3 & 0 & 0 \\ 0 & 2 & 0 \\ 0 & 0 & 1 \end{bmatrix}, \quad F_1 = \begin{bmatrix} 1 & 0 & 0 \\ 0 & \frac{1}{2} & 0 \\ 0 & 0 & 0 \end{bmatrix},$$

$$F_2 = \begin{bmatrix} 2 & 0 & 0 \\ 0 & 1 & 0 \\ 0 & 0 & 0 \end{bmatrix} \quad \text{and} \quad \theta = 0.$$

Let $p'\hat{\beta}_1$ and $p'\hat{\beta}_2$ be the BE derived from F_1 and F_2. Find the parallelepiped where the MSE of $p'\hat{\beta}_1$ is smaller than that of $p'\hat{\beta}_2$.

Exercise 7.5.10. Give the form of the s dimensional rectangle in (7.5.19):

A. for comparing ridge, generalized ridge contraction and generalized contraction estimators with each other;

B. for comparing individual, total or predictive MSE of a BE and a LS estimator.

7.6 Summary

The MSE of the Bayes estimator and the mixed estimator neglecting prior assumptions has been calculated. Comparisons of the MSE were made:

1. between the Bayes or mixed estimator and the least square estimator;

2. between Bayes estimators.

The regions where the BE or mixed estimators were optimal with respect to the least square estimator were ellipsoids about the prior mean.

Mean square errors averaging over quadratic loss functions were considered and similar comparisons resulted. The problem of comparing two Bayes estimators was considerably simplified.

The general results were applied to compare ridge type estimators to one another.

The comparison of the MSE for estimators derived from incorrect prior assumptions averaged over the correct prior assumptions is a related subject to be investigated in the next chapter.

Chapter VIII

The MSE for Incorrect Prior Assumptions

8.0 Introduction

A strong, serious but valid criticism of Bayesian methods is that the prior information they depend on may not be correct. An important question, therefore is: How robust is an estimator with respect to a prior distribution? To what extent is an estimator still a good one when the prior information is incorrect or questionable?

In the context of the development of the estimators in this book the form of the incorrect prior assumptions for the Bayes, mixed and minimax estimators are respectively:

1. an incorrectly specified mean vector and/or incorrectly specified dispersion matrix;
2. incorrect stochastic prior information;
3. an incorrect ellipsoid.

To study the robustness of the estimators with respect to prior assumptions the following questions should be answered.

1. What is the MSE of an estimator derived from incorrect prior assumptions averaging over correct prior assumptions?
2. How does this MSE compare with that of the least square estimator?

3. How do the mathematical form and properties of this MSE compare with the average and conditional MSE of Chapters VI and VII?

The MSE of the BE is obtained in Section 8.1. It is less than the MSE in Chapter VI when the incorrect prior mean lies in an ellipsoid containing the correct prior mean (see Theorem 8.1.1). The MSE of an estimator derived from the incorrect prior assumptions is not uniformly less than that of the least square estimator when averaged over correct prior assumptions (see Theorem 8.1.2 and its corollary). In Chapter VI the average MSE was uniformly less than that of the LS estimator.

The corresponding comparisons are made in Section 8.2 for the minimax estimator.

Section 8.3 considers:

1. the corresponding comparisons of the MSE of the mixed estimator to the LS estimator when the mean and dispersion of ϕ is misspecified (Judge (1978) calls this biased prior information);
2. the correspondence between incorrect prior assumptions for a BE and a mixed estimator.

Of course the correct prior distribution may not be known. Maybe information is available about the relative weight one would give to one prior as compared to another. Section 8.4 considers estimators and their properties derived from mixtures of two prior distributions. The ideas considered there are similar to those of Berger and Berliner (1986).

Another possibility would be to choose between two different Bayes estimators for different prior distributions with a given probability. The MSE properties of these estimators are taken up in Section 8.5.

8.1 The BE and Its MSE

Consider the BE derived for incorrect prior assumptions. Suppose the MSE of this BE is calculated averaging over the correct prior assumptions? The following questions suggest themselves.

1. Under what conditions is the MSE less than that obtained averaging over the original incorrect prior assumptions?
2. When is the MSE less than that of the LS estimator?

   To answer these questions first the MSE of the BE (derived from incorrect assumptions) must be calculated averaging over the correct prior assumptions. The answers to the questions are covered in Theorems 8.1.1 and 8.1.2 and its corollaries below. Whether the estimators are optimal for this MSE depends on the degree of "closeness" of the two sets of prior assumptions. When only the prior mean is incorrectly specified its MSE is less than that of LS when θ lies in an ellipsoid about η.

   Notice that the MSE of the BE that is calculated in this section will be less than that of the LS estimator only for a range of values of the θ parameter. The MSE averaging over the prior in Chapter VI was smaller than LS and did not depend on the θ parameter.

   Suppose that the prior assumptions

$$E(\beta) = \theta \quad \text{and} \quad D(\beta) = F \tag{8.1.1}$$

are incorrect. The correct prior assumptions are

$$E(\beta) = \eta \quad \text{and} \quad D(\beta) = W. \tag{8.1.2}$$

The BE is obtained with respect to (8.1.1). Since (8.1.1) is incorrect the MSE is calculated averaging over (8.1.2).

   Three forms of the MSE may be obtained from (7.1.6), (7.1.9) and (7.1.12) by calculating their expectations averaging over (8.1.2). Notice that

$$E(\beta - \theta)(\beta - \theta)' = W + (\theta - \eta)(\theta - \eta)'. \tag{8.1.3}$$

Let

$$C_1 = UU'FUU + \sigma^2(X'X)^+$$

and

$$\Gamma_1 = X'X + \sigma^2(UU'FUU')^+.$$

From (7.1.6) and (8.1.3), the first form of the MSE is

$$
\begin{aligned}
m_3 =\; & p'\Gamma_1^+ X'X\Gamma_1^+ p\sigma^2 \\
& + p'\Gamma_1^+(UU'FUU')^+(W + (\theta - \eta)(\theta - \eta)')(UU'FUU')^+\Gamma_1^+ p.
\end{aligned}
\tag{8.1.4}
$$

From (7.1.9) and (8.1.3), the second form of the MSE is

$$
\begin{aligned}
m_3 =\; & p'FC_1^+ X'XFp \\
& + p'[I - FC_1^+][W + (\theta - \eta)(\theta - \eta)'][I - C_1^+ F]p.
\end{aligned}
\tag{8.1.5}
$$

From (7.1.12) and (8.2.3), the third form of the MSE is

$$
\begin{aligned}
m_3 =\; & p'(X'X)^+ p\sigma^2 - p'(X'X)^+ C_1^+(X'X)^+ p\sigma^4 \\
& + p'(X'X)^+ C_1^+[(X'X)^+\sigma^2 + W + (\theta - \eta)(\theta - \eta)'] \\
& \times C_1^+(X'X)^+ p\sigma^4.
\end{aligned}
\tag{8.1.6}
$$

The following theorem obtains the form of the region about θ where the MSE of the BE averaging over the original "correct" prior assumptions is smaller than that obtained averaging over the original "incorrect" prior assumptions.

Theorem 8.1.1. *If either of the following conditions hold:*

(1) $W + (\theta - \eta)(\theta - \eta)' \leq F$ \hfill (8.1.7)

or

(2) $W \leq F, \theta - \eta$ *belongs to the range of* $F - W$ *and*
  $(\theta - \eta)'(F - W)^+(\theta - \eta) \leq 1,$ \hfill (8.1.8)

the MSE for the correct prior assumptions (8.1.4)–(8.1.6) *is less than the MSE for the incorrect prior assumptions* (6.3.9)–(6.3.11).

If $W \geq F$ *the MSE for the correct prior assumptions is greater than that for the incorrect prior assumptions.*

Proof. Suppose condition (1) holds. Then

$$(UU'FUU')^+(W + (\theta - \eta)(\theta - \eta)')(UU'FUU')^+$$

$$\leq (UU'FUU')^+F(UU'FUU')^+$$

$$= U(U'FU)^{-1}U'FU(U'FU)^{-1}U' \qquad (8.1.9)$$

$$= U(U'FU)^{-1}U' = (UU'FUU')^+.$$

Then, from (8.1.9)

$$p'\Gamma_1^+ X'X\Gamma_1^+ p$$

$$+ p'\Gamma_1^+(UU'FUU')^+(W + (\theta - \eta)(\theta - \eta)')$$

$$\times (UU'FUU')^+\Gamma_1^+ p\sigma^4 \qquad (8.1.10)$$

$$\leq p'\Gamma_1^+\Gamma_1\Gamma_1^+ p\sigma^2 = p'\Gamma_1^+ p\sigma^2.$$

The right hand side of inequality (8.1.10) is the MSE averaging over the incorrect prior assumptions.

Suppose condition (2) holds. From Theorem 2.5.3

$$(\theta - \eta)(\theta - \eta)' + W \leq F. \qquad (8.1.11)$$

The result then follows from part 1.

When $W \geq F$ the inequality in (8.1.11) can be reversed. ∎

The next result will enable comparison of the MSE of the BE with OLS averaging over the correct prior assumptions, given that the original prior assumptions are incorrect. An ellipsoid about the correct prior mean η is obtained.

Theorem 8.1.2. *Let N be any NND matrix. Suppose a BE $p'\hat{\beta}$ was derived with respect to incorrect prior assumptions (8.1.1). Then averaging over the correct prior assumptions (8.1.2).*

$$m_3 \le p'Np \quad \text{iff} \quad (\theta - \eta)'T^+(\theta - \eta) < 1 \qquad (8.1.12)$$

with

$$T = F\Gamma_1 N\Gamma_1 F\frac{1}{\sigma^4} - FX'XF\frac{1}{\sigma^2} - W \qquad (8.1.13)$$

and $U'TU$ is PD.

Proof. From (8.1.4) inequality (8.1.12) holds iff

$$U'[\sigma^2 X'X + (UU'FUU')^+[W + (\theta - \eta)(\theta - \eta)'](UU'FUU')^+\sigma^4]U$$

$$\le (\Lambda + \sigma^2(U'FU)^{-1})N(\Lambda + \sigma^2(U'FU)^{-1}) \qquad (8.1.14)$$

or, equivalently,

$$U'[W + (\theta - \eta)(\theta - \eta)']U\sigma^4$$

$$\le (U'FU)(\Lambda + \sigma^2(U'FU)^{-1})U'NU(\Lambda + \sigma^2(U'FU)^{-1})U'FU$$

$$- \sigma^2 U'FX'XFU. \qquad (8.1.15)$$

Inequality (8.1.15) holds true iff

$$U'(\theta - \eta)(\theta - \eta)U' \le U'TU \qquad (8.1.16)$$

with T defined in (8.1.14) above. Inequality (8.1.16) holds true iff (8.1.12) holds by Theorem 2.8.2. ∎

The following corollary gives the comparison of the MSE of the BE with that of OLS.

Corollary 8.1.1. *The MSE*

$$m_3 \leq \text{MSE}(p'b_1) \tag{8.1.17}$$

iff

(1) $2U'FU + \sigma^2\Lambda^{-1} - U'WU > 0$ (8.1.18)

and

(2) θ *lies in the ellipsoid* (8.1.12).

When $F = W$ *condition* (2) *becomes*

$$(\theta - \eta)'[\sigma^2(X'X)^+ + UU'FUU']^+(\theta - \eta) < 1. \tag{8.1.19}$$

Proof. Since

$$\text{MSE}(p'b) = \sigma^2 p'(X'X)^+ p, \tag{8.1.20}$$

let $N = \sigma^2(X'X)^+$. Then

$$T = F\Gamma_1(X'X)^+\Gamma_1 F\frac{1}{\sigma^2} - \frac{1}{\sigma^2}FX'XF - W. \tag{8.1.21}$$

Now

$\Gamma_1(X'X)^+\Gamma_1$

$$= [X'X + \sigma^2 U(U'FU)^{-1}U'](X'X)^+[X'X + \sigma^2 U(U'FU)^{-1}U']$$

$$= X'X + 2\sigma^2 U(U'FU)^{-1}U' + \sigma^4 U(U'FU)^{-1}\Lambda^{-1}(U'FU)^{-1}U'. \tag{8.1.22}$$

Then, by (8.1.2) and (8.1.3),

$$U'TU = 2U'FU + \sigma^2\Lambda^{-1} - U'WU. \tag{8.1.23}$$

Thus $U'TU > 0$ is equivalent to (8.1.18). In fact $T > 0$ implies (8.1.18). Then when θ lies in (8.1.12), (8.1.8) holds.

When $F = W$,

$$UTU' = U'FU + \sigma^2 \Lambda^{-1}. \tag{8.1.24}$$

Equation (8.1.19) follows. ∎

The result in (8.1.19) is similar to that of (7.3.8).

The corresponding results for the minimax estimator will be given after the examples below.

Example 8.1.1. Comparison of Ridge Type Estimators Under Incorrect Prior Assumptions. The contraction estimator

$$p'\hat{\beta} = \frac{1}{1+k}p'b$$

is derived from the prior assumptions

$$E(\beta) = 0, \quad D(\beta) = \frac{\sigma^2}{k}(X'X)^+. \tag{8.1.25}$$

Suppose the prior assumptions in (8.1.25) are incorrect and that the correct prior assumptions are

$$E(\beta) = \theta \quad \text{and} \quad D(\beta) = \frac{\sigma^2}{l}(X'X)^+. \tag{8.1.26}$$

Then

1. If $k < l$, there is a vector a such that $(X'X)a = \theta$ and

$$\theta'(X'X)\theta < \frac{\sigma^2 kl}{l-k};$$

then by Theorem 8.1.1 the contraction estimator has a smaller MSE averaging over (8.1.26) than (8.1.25).

2. If

$$l > \frac{k}{k+2}$$

and θ lies in the ellipsoid

$$\theta'(X'X)\theta < \frac{\sigma^2 kl}{kl + 2l - k}$$

then by Corollary 8.1.1 the MSE of the contraction estimator of $p'\beta$ averaging over (8.1.26) is less than that of the LS.

Exercise 8.1.1. Derive the explicit conditions of Theorems 8.1.1 and 8.1.2 and Corollary 8.1.1:

A. for the ridge estimator;
B. for a BE with nonzero prior mean and dispersion $(\sigma^2/k)I$.

Exercise 8.1.2. What would be the explicit conditions of Theorems 8.1.1 and 8.1.2 for comparing:

A. two contraction estimators?
B. two ordinary ridge estimators?
C. two generalized ridge estimators?
D. two generalized contraction estimators?

Exercise 8.1.3. Consider the model in Exercise 3.5.2 where the prior assumptions are incorrect. The correct prior assumptions are

$$E(\mu) = \eta, \quad D(\mu) = w^2.$$

A. From Theorem 8.1.1 what are the conditions for the MSE for the correct prior assumptions to be less than that for the incorrect prior assumptions?
B. What are the conditions for the MSE of this estimator with the correct prior assumptions to be less than that of the OLS (Corollary 8.1.1)?

Exercise 8.1.4. For the linear model (3.1.1) consider the different ridge type estimators whose prior dispersion is of the form UDU' (i.e. ridge, generalized ridge, contraction, etc.). Suppose instead of 0, $\eta \neq 0$ is the correct prior mean and instead of the prior for one type of ridge estimator the prior for another kind should be used (Example: a contraction estimator could be used in place of an ordinary ridge estimator). Write down the inequalities of Theorem 8.1.1 and 8.1.2 for the different comparisons. Then repeat the exercise when $\eta = 0$.

8.2 The Minimax Estimator

In Section 3.4 the minimax estimator was defined as an estimator of the form

$$p'\hat{\beta} = p'\theta + L'(Y - X\theta), \qquad (8.2.1)$$

where the maximum MSE on an ellipsoid of the form

$$(\beta - \theta)'G(\beta - \theta) \leq 1 \qquad (8.2.2)$$

is minimized. The form of the resulting estimator was the same as that of the BE with $G^+ = F$. Thus, (8.2.2) corresponds to the prior assumptions

$$E(\beta) = \theta \quad \text{and} \quad D(\beta) = G^+ \qquad (8.2.3)$$

for the BE.

By analogy with Section 8.1 suppose the minimax estimator is obtained assuming

$$(\beta - \theta)'F^+(\beta - \theta) \leq 1, \qquad (8.2.4a)$$

but the ellipsoid in (8.2.4a) is incorrect. Suppose the correct ellipsoid is

$$(\beta - \eta)'W^+(\beta - \eta) \leq 1. \qquad (8.2.4b)$$

Then, given the conditions of Theorems 8.1.1 and 8.1.2:

1. The maximum MSE on (8.2.4b) is less than that on (8.2.4a).
2. The maximum MSE on (8.2.4b) is less than that of the OLS if the conditions of Corollary 8.1.2 hold true.

In summary the maximum MSE of the minimax estimator derived using (8.2.1) on (8.2.4b) has the same properties as the MSE of a BE averaged over the correct prior assumptions. Similar results can be obtained for the mixed estimator.

Example 8.2.1. Minimax Estimator's Efficiency on Different Ellipsoids. Suppose that for the linear model in Example 6.5.1 the minimax estimator is derived by minimizing the maximum MSE of $p'\beta$ on the ellipsoid

$$\frac{\gamma_1^2}{4\sigma^2} + \frac{\gamma_2^2}{2\sigma^2} \leq 1.$$

If the correct ellipsoid should have been

$$\frac{\gamma_1^2}{2\sigma^2} + \frac{\gamma_2^2}{\sigma^2} \leq 1,$$

i.e., the parameters should have been restricted more, the minimum value of the maximum MSE is smaller for the smaller ellipsoid.

If the correct smaller ellipsoid is not centered at the origin but on $\eta \neq 0$ then if

$$U'\eta\eta'U \leq \begin{bmatrix} 2 & 0 \\ 0 & 1 \end{bmatrix} \sigma^2,$$

i.e., $\xi = U'\eta$ lies on the ellipsoid

$$\frac{\xi_1^2}{2\sigma^2} + \frac{\xi_2^2}{\sigma^2} \leq 1,$$

the minimum of the maximum MSE is smaller. Here $W = F_1$ and $F = F_2$ in Example 7.3.3.

Exercise 8.2.1. For the above example

A. Give the precise matrices and ellipsoids for conditions 1 and 2.
B. When $F = W = F_2$ and $\theta = 0$ give the precise ellipsoid (8.1.20).

8.3 The Mixed Estimator

In Section 8.1 the MSE of a BE derived from an incorrect prior mean and variance was considered. In Section 8.2 the maximum value on the correct ellipsoid of a minimax estimator derived by finding the optimum estimator for an incorrect ellipsoid was obtained. Analogously, for the augmented model the mixed estimator can be derived given incorrect stochastic prior information. However, the variance can be calculated as if the prior information was correct.

Consider the augmented model

$$\begin{bmatrix} Y \\ r \end{bmatrix} = \begin{bmatrix} X \\ R \end{bmatrix} \beta + \begin{bmatrix} \epsilon \\ \phi \end{bmatrix} \tag{8.3.1}$$

together with the assumptions

$$E(\epsilon) = 0 \quad \text{and} \quad D(\epsilon) = \sigma^2 I, \tag{8.3.2}$$

$$E(\phi) = 0 \quad \text{and} \quad D(\phi) = H, \tag{8.3.3}$$

where H is a nonsingular matrix. Assume that the assumptions in (8.3.3) are incorrect. Suppose that the correct assumptions are

$$E(\phi) = \delta \quad \text{and} \quad D(\phi) = Q. \tag{8.3.4}$$

When $\delta \neq 0$ the prior assumption is said to be biased because the mixed estimator

$$p'\hat{\beta}_c = p'(X'X + \sigma^2 R'H^{-1}R)^+(X'Y + \sigma^2 R'H^{-1}r) \tag{8.3.5}$$

has expectation, for $p'\beta$ estimable,

$$E(p'\hat{\beta}_c) = p'(X'X + \sigma^2 R'H^{-1}R)^+(X'X + \sigma^2 R'H^{-1}R)\beta$$

$$+ p'(X'X + \sigma^2 R'H^{-1}R)^+ R'H^{-1}\delta\sigma^2. \tag{8.3.6}$$

For the assumptions (8.3.4) the MSE of (8.3.5) is

$$\mathrm{MSE}(p'\hat{\beta}_c) = p'\Gamma_2^+ X'X\Gamma_2^+ p\sigma^2 + p'\Gamma_2^+ R'H^{-1}(Q + \delta\delta')H^{-1}R\Gamma_2^+ p\sigma^4 \tag{8.3.7}$$

with $\Gamma_2 = X'X + (R'H^{-1}R)\sigma^2$.

Two questions suggest themselves:

1. When is the MSE in (8.3.7) the same as that of the BE?
2. When is the MSE in (8.3.7) less than that of the LS estimator?

The first question is answered by observing that if

$$\delta = R(\theta - \eta), \quad (UU'FUU')^+ = R'H^{-1}R \tag{8.3.8}$$

and

$$Q = RWR', \tag{8.3.9}$$

equations (8.3.7) and (8.1.4) are equivalent.

The second question is answered by Theorem 8.3.1 below.

Theorem 8.3.1. *If either*

1. $\delta = 0$, $R = AU'$ and
$$2H + \sigma^2 A\Lambda^{-1}A - Q > 0 \tag{8.3.10}$$

or

2. $\delta \neq 0$, $Q = H$ and
$$\delta'(H + \sigma^2 R(X'X)^+ R')^+\delta \leq 1 \tag{8.3.11}$$

then

$$\mathrm{MSE}(p'\hat{\beta}_c) \leq \mathrm{MSE}(p'b). \tag{8.3.12}$$

Proof. When $\delta = 0$ and $p'\hat{\beta}_c$ is estimable (8.3.12) holds true if

$$(\Lambda + \sigma^2 AH^{-1}A)^{-1}[\sigma^2\Lambda + \sigma^4 AH^{-1}QH^{-1}A](\Lambda + \sigma^2 AH^{-1}A)^{-1}$$
$$\leq \sigma^2\Lambda^{-1}, \tag{8.3.13}$$

or

$$2\sigma^4 AH^{-1}A + \sigma^6 AH^{-1}A\Lambda^{-1}AH^{-1}A - \sigma^4 AH^{-1}QH^{-1}A \geq 0, \tag{8.3.14}$$

or

$$2H + \sigma^2 A\Lambda^{-1} A - Q \geq 0. \qquad (8.3.15)$$

When $\delta \neq 0$ and $Q = H$, (8.3.12) holds true if

$$(\Lambda + \sigma^2 A H^{-1} A)^{-1} A H^{-1} \delta\delta' H^{-1} A (\Lambda + \sigma^2 A H^{-1} A)^{-1} \sigma^4$$

$$\leq \sigma^2 \Lambda^{-1} - (\Lambda + \sigma^2 A H^{-1} A)^{-1} \Lambda (\Lambda + \sigma^2 A H^{-1} A)^{-1} \sigma^2$$

$$- (\Lambda + \sigma^2 A H^{-1} A)^{-1} A H^{-1} A (\Lambda + \sigma^2 A H^{-1} A)^{-1} \sigma^4, \quad (8.3.16)$$

or

$$\sigma^4 A H^{-1} \delta\delta' H^{-1} A$$

$$\leq \sigma^2 (\Lambda + \sigma^2 A H^{-1} A) \Lambda^{-1} (\Lambda + \sigma^2 A H^{-1} A) - \sigma^2 \Lambda - \sigma^4 A H^{-1} A$$

$$= \sigma^4 A H^{-1} A + \sigma^6 A H^{-1} A \Lambda^{-1} A H^{-1} A. \qquad (8.3.17)$$

From (8.3.17)

$$\delta\delta' \leq H + A\Lambda^{-1} A \sigma^2, \qquad (8.3.18)$$

or, by Theorem 2.5.2,

$$\delta'(H + A\Lambda^{-1} A \sigma^2)^{-1} \delta = \delta'(H + R(X'X)^+ R')^+ \delta \leq 1. \blacksquare \quad (8.3.19)$$

Of course prior assumptions may not be really known. This situation will be dealt with in Sections 8.4 and 8.5.

Example 8.3.1. Comparison of Mixed Estimator with LS Under Incorrect Prior Assumptions. Consider the model in Example 3.1.1. Let

$$r = \begin{bmatrix} 1 & 1 & 0 \\ 0 & 1 & 1 \end{bmatrix} \begin{bmatrix} \beta_0 \\ \beta_1 \\ \beta_2 \end{bmatrix} + \phi,$$

$$E(\phi) = \begin{bmatrix} \delta_1 \\ \delta_2 \end{bmatrix}, \quad D(\phi) = \tau^2 I \text{ and } \sigma^2 = 1.$$

The condition in (8.3.11) is

$$[\delta_1 \quad \delta_2] \left(I + \begin{bmatrix} 1 & 1 & 0 \\ 0 & 1 & 1 \end{bmatrix} \begin{bmatrix} \frac{1}{6} & \frac{1}{6} & 0 \\ \frac{1}{6} & \frac{1}{6} & 0 \\ 0 & 0 & 1 \end{bmatrix} \begin{bmatrix} 1 & 0 \\ 1 & 1 \\ 0 & 1 \end{bmatrix} \right)^{-1} \begin{bmatrix} \delta_1 \\ \delta_2 \end{bmatrix} \leq 1$$

or

$$[\delta_1 \ \delta_2] \begin{bmatrix} \frac{12}{19} & -\frac{3}{19} \\ -\frac{3}{19} & \frac{15}{19} \end{bmatrix} \begin{bmatrix} \delta_1 \\ \delta_2 \end{bmatrix} \leq 1. \qquad (8.3.20)$$

Exercise 8.3.1. Find a transformation so that the quadratic form in (8.3.20) has a diagonal matrix. What are the axes of the ellipse?

Exercise 8.3.2. Does $\delta = \begin{bmatrix} 1 \\ -1 \end{bmatrix}$ satisfy (8.3.20)?

Exercise 8.3.3. When do the conditions of Theorem 8.3.2 hold if

A. $X = \begin{bmatrix} 1_3 & 1_3 & 0 & 0 \\ 1_3 & 0 & 1_3 & 0 \\ 1_3 & 0 & 0 & 1_3 \end{bmatrix}$, $\quad R = \begin{bmatrix} 1 & 1 & 1 & 1 \\ 1 & -1 & 0 & 0 \\ 1 & 1 & -2 & 0 \\ 1 & 1 & 1 & -3 \end{bmatrix}$,

$Q = I$, $\quad H = I$?

B. The design matrix X is as in A above, $X = R$, $Q = H = I$?

8.4 Contaminated Priors

From a practical standpoint prior information is useful only when it is correct. Perhaps knowledge is available about the probability of correctness of two or more priors. In that situation, perhaps a mixture of these priors is appropriate. The Bayes estimator with respect to such a prior may then be derived by the method of Section 3.5.

Questions that are worth investigating are:

1. When is the MSE of this BE less than that of the ordinary least square estimator?
2. How does its MSE compare with that of the BE derived from one of the priors in the mixture?
3. How does the BE derived from a prior that is a component of the mixture perform when averaged over the mixed distribution?

Consider the two prior distributions $\pi_1(\beta)$ and $\pi_2(\beta)$ with means θ_1 and θ_2 and dispersions F_1 and F_2. Suppose $\pi_1(\beta)$ is the correct prior with probability ε_1 and $\pi_2(\beta)$ is the correct prior with probability $1 - \varepsilon_1$. The distribution

$$\pi_{\varepsilon_1}(\beta) = \varepsilon_1 \pi_1(\beta) + (1 - \varepsilon_1)\pi_2(\beta) \tag{8.4.1}$$

is called the ε contaminated prior. The mean and the dispersion are respectively

$$\theta_{\varepsilon_1} = \varepsilon_1 \theta_1 + (1 - \varepsilon_1)\theta_2 \tag{8.4.2a}$$

and

$$F_{\varepsilon_1} = \varepsilon_1 F_1 + (1 - \varepsilon_1)F_2. \qquad (8.4.2b)$$

The BE for the prior information (8.4.1) and (8.4.2) is

$$p'\hat{\beta}\varepsilon_1 = p'\theta_{\varepsilon_1} + p'F_{\varepsilon_1}X'(XF_{\varepsilon_1}X' + \sigma^2 I)^{-1}(Y - X\theta_{\varepsilon_1}). \qquad (8.4.3)$$

Let $p'\hat{\beta}\varepsilon_1$ and $p'\hat{\beta}\varepsilon_2$ be the BE derived from priors with dispersions F_1 and F_2 respectively.

Theorem 8.4.1 below explains that if $F_1 \leq F_2$ the MSE of $p'\hat{\beta}_\varepsilon$ is somewhere in between that of $p'\hat{\beta}_1$ and $p'\hat{\beta}_2$. Theorem 8.4.2 gives the ellipsoid where $p'\hat{\beta}_\varepsilon$ has a smaller conditional MSE than LS. A comparison of the average and conditional MSE is made in Theorem 8.4.3. Theorem 8.4.4 compares the MSE of $p'\hat{\beta}_1$ and $p'\hat{\beta}_2$ averaging over $\pi_{\varepsilon_1}(\beta)$ with that obtained averaging over $\pi_1(\beta)$ and $\pi_2(\beta)$.

Theorem 8.4.1. *Let $p'\hat{\beta}_1$ be the BE assuming the prior mean θ_1 and dispersion F_1, $p'\hat{\beta}_2$ be the BE assuming the prior mean θ_2 and dispersion F_2 and $p'\hat{\beta}_\varepsilon$ the BE with respect to the contaminated prior. Then averaging over the respective priors*

$$\mathrm{MSE}(p'\hat{\beta}_1) \leq \mathrm{MSE}(p'\hat{\beta}_{\varepsilon_1}) \leq \mathrm{MSE}(p'\hat{\beta}_2). \qquad (8.4.4)$$

Proof. The result follows from Theorem 6.5.1 because $F_1 \leq F_2$ implies

$$F_1 = \varepsilon_1 F_1 + (1 - \varepsilon_1)F_1 \leq \varepsilon_1 F_1 + (1 - \varepsilon_1)F_2$$
$$\leq \varepsilon_1 F_2 + (1 - \varepsilon_1)F_2 = F_2. \blacksquare$$

The average MSE of $p'\hat{\beta}_{\varepsilon_1}$ is less than that of the least square estimator by (6.3.1). A simple consequence of Theorem 7.3.1, Corollary 7.3.1 is

Theorem 8.4.2. *The conditional MSE of $p'\hat{\beta}\varepsilon_1$ is less then that of the LS $p'b$*

$$(\beta - \theta_{\varepsilon_1})'(2UU'F_{\varepsilon_1}UU' + \sigma^2(X'X)^+)^+(\beta - \theta_{\varepsilon_1}) \leq 1. \qquad (8.4.5)$$

If $\varepsilon_1 = 1$ (8.4.5) reduces to the condition that $p'\hat{\beta}_1$ have a smaller MSE than LS. If $\varepsilon_1 = 0$ (8.4.5) reduces to the condition that $p'\hat{\beta}_2$ have a smaller MSE than LS.

Also, from Corollary 7.3.2,

Theorem 8.4.3. *The conditional MSE of $p'\hat{\beta}_{\varepsilon_1}$ is less than the average MSE if*

$$(\beta - \theta_{\varepsilon_1})(UU'F_{\varepsilon_1}UU')^+(\beta - \theta_{\varepsilon_1}) \leq 1. \qquad (8.4.6)$$

Inequality (8.4.6) reduces to the corresponding comparison for $p'\hat{\beta}_1$ when $\varepsilon_1 = 1$ and for $p'\hat{\beta}_2$ when $\varepsilon_1 = 0$.

When $F_1 \leq F_2$ the following comparison of the MSE of $p'\hat{\beta}_1$ and $p'\hat{\beta}_2$ averaging over $\pi_1(\beta)$ and $\pi_2(\beta)$ respectively may be made with that averaging over $\pi_\varepsilon(\beta)$ as a consequence of Theorem 8.1.1.

Theorem 8.4.4.
(1) Let $F_1 \leq F_2$. If $\theta_1 - \theta_2$ belongs to the range of $F_2 - F_1$ and

$$(\theta_1 - \theta_2)'(F_2 - F_1)^+(\theta_1 - \theta_2) \leq 1/(1 - \varepsilon_1) \qquad (8.4.7)$$

the MSE of $p'\hat{\beta}_2$ averaging over $\pi_\varepsilon(\beta)$ is less than that averaging over $\pi_2(\beta)$.
(2) The MSE of $p'\hat{\beta}_1$ averaging over $\pi_\varepsilon(\beta)$ is greater than the MSE averaging over $\pi_1(\beta)$.

A variation of the problem considered above will now be considered.

Example 8.4.1. Ridge Estimator with Contaminated Prior.
Consider a contaminated prior with means zero and

$$F_\epsilon = \epsilon \frac{\sigma^2 I}{k_0} + (1-\epsilon)\frac{\sigma^2 I}{k_1}.$$

The resulting estimator is a ridge estimator with

$$\frac{\sigma^2}{k_\epsilon} = \frac{\epsilon \sigma^2}{k_0} + \frac{(1-\epsilon)\sigma^2}{k_1}$$

so

$$\frac{1}{k_\epsilon} = \frac{\epsilon k_1 + (1-\epsilon)k_0}{k_0 k_1}$$

and

$$k_\epsilon = \frac{k_0 k_1}{\epsilon k_1 + (1-\epsilon)k_0}.$$

Exercise 8.4.1. Let K_0 and K_1 be diagonal matrices,

$$F_\epsilon = \epsilon \sigma^2 U K_0^{-1} U' + (1-\epsilon)\sigma^2 U K_1^{-1} U'.$$

The prior means are zero. What is the form of the matrix K and
what are its elements in terms of those of K_0 and K_1 for the resulting
generalized ridge estimator?

Exercise 8.4.2. Suppose that the parameter u in the linear model

$$y_i = \mu + \epsilon_i, \qquad 1 \leq i \leq n,$$

has prior distribution

$$\pi(\mu) = \int_A \pi_\lambda(\mu) \, dF(\mu),$$

where A is a subset of R' (real line) and the integral is Lebesgue-Stieltjes, i.e., $\pi_\lambda(\mu)$ is a mixture of different priors for different λ. (If $A = [0,1]$, this reduces to $\epsilon\pi_0(\mu) + (1 - \epsilon)\pi_1(\mu)$.) Obtain expressions for the prior mean and variance of μ stating carefully any assumptions you make.

Exercise 8.4.3. Formulate the optimization problem for the minimax estimator that leads to (8.4.4).

Exercise 8.4.4. What is the MSE of the estimator in Exercise 8.4.1?

8.5 Contaminated (Mixed) Bayes Estimators

Suppose that there are two Bayes estimators $p'\hat{\beta}_1$ and $p'\hat{\beta}_2$ derived from prior distributions with means and variances θ_1, F_1 and θ_2, F_2 respectively. It is believed that $p'\hat{\beta}_1$ is the better or correct estimator with probability ε_2 and $p'\hat{\beta}_2$ is the correct estimator with probability $1 - \varepsilon_2$. Such an estimator can be obtained by minimizing

$$ m = \varepsilon_2 \, \text{Var}(p'\beta - a_{01} - L_1'Y) + (1-\varepsilon_2) \, \text{Var}((p'\beta - a_{02} - L_2'Y) \quad (8.5.1) $$

given

$$ E(p'\beta - a_{01} - L_1'Y) = 0 \quad \text{and} \quad E(p'\beta - a_{02} - L_2'Y) = 0, \quad (8.5.2) $$

where the first term in (8.5.1) is averaged over π_1 and the second term is averaged over π_2. The resulting estimator $p'\hat{\beta}_{\varepsilon_2}$ is $p'\hat{\beta}_1$ with probabailty ε_2 and $p'\hat{\beta}_2$ with probability $1 - \varepsilon_2$. Its average MSE is given by

$$ \text{MSE}(p'\hat{\beta}\varepsilon_2) = \varepsilon_2 \, \text{MSE}(p'\hat{\beta}_1) + (1 - \varepsilon_2) \, \text{MSE}(p'\hat{\beta}_2). \quad (8.5.3) $$

Thus, a form of the average MSE is

$$
\begin{aligned}
\mathrm{MSE}(p'&\hat{\beta}\varepsilon_2) \\
&= p'(X'X)^+ p\sigma^2 \\
&\quad - \varepsilon_2 p'(X'X)^+ (UU'F_1 UU' + \sigma^2(X'X)^+)^+ (X'X)^+ p\sigma^4 \\
&\quad - (1-\varepsilon_2) p'(X'X)^+ (UU'F_2 UU' + \sigma^2(X'X)^+)^+ (X'X)^+ p\sigma^4.
\end{aligned}
$$
$$(8.5.4)$$

However, this MSE is larger than that of $p'\hat{\beta}_{\varepsilon_1}$. This is shown by Theorem 8.5.1 below.

Theorem 8.5.1. *If $U'F_1 U$ and $U'F_2 U$ are PD, the average MSE of $p'\hat{\beta}_{\varepsilon_1}$ and $p'\hat{\beta}_{\varepsilon_2}$ satisfy*

$$
\mathrm{MSE}(p'\hat{\beta}_{\varepsilon_1}) \le \mathrm{MSE}(p'\hat{\beta}_{\varepsilon_2}). \tag{8.5.5}
$$

Proof. From the properties of estimability and the SVD of the matrices involved it is enough to show

$$
(U'F_{\varepsilon_2} U + \sigma^2\Lambda^{-1})^{-1} \le (U'F_1 U + \sigma^2\Lambda^{-1})^{-1}\varepsilon_2
$$

$$
+ (U'F_2 U + \sigma^2\Lambda^{-1})^{-1}(1-\varepsilon_2). \tag{8.5.6}
$$

Let $A_1 = U'F_1 U + \sigma^2\Lambda^{-1}$ and $A_2 = U'F_2 U + \sigma^2\Lambda^{-1}$. Rewrite (8.5.6) as

$$
(\varepsilon_2 A_1 + (1-\varepsilon_2)A_2)^{-1} \le A_1^{-1}\varepsilon_2 + A_2^{-1}(1-\varepsilon). \tag{8.5.7}
$$

From C.R. Rao (1973) there is a matrix R and a diagonal matrix Δ such that

$$
A_1 = R'^{-1}\Delta R^{-1} \quad \text{and} \quad A_2 = R'^{-1}R^{-1}. \tag{8.5.8}
$$

Then (8.5.7) reduces to

$$
(\varepsilon_2\Delta + (1-\varepsilon_2)I)^{-1} \le \Delta^{-1}\varepsilon_2 + (1-\varepsilon_2)I. \tag{8.5.9}
$$

Let δ_1 be the elements of Δ. To establish (8.5.9) it suffices to show

$$\frac{1}{\varepsilon_2 \delta_1 + (1 - \varepsilon_2)} \leq \frac{\varepsilon_2}{\delta_1} + 1 - \varepsilon_2. \tag{8.5.10}$$

Equation (8.5.10) holds true iff

$$\delta_1 \leq [\varepsilon_2 \delta_1 + (1 - \varepsilon_2)][\varepsilon_2 + (1 - \varepsilon_2)\delta_1]. \tag{8.5.11}$$

Simple algebra shows that (8.5.11) reduces to

$$0 \leq (\delta_1 - 1)^2. \tag{8.5.12}$$

Thus (8.5.10) holds true. This implies (8.5.9) which in turn implies (8.5.7) and (8.5.6). The result then follows. ∎

Clearly the average MSE of $p'\hat{\beta}\epsilon$ is less than that of the LS. For the conditional MSE the following statement may be made.

Theorem 8.5.2. *If* $F_1 \leq F_2$ *and*

$$(\beta - \theta)'(UU'F_1UU' + \sigma^2(X'X)^+) + (\beta - \theta) \leq 1 \tag{8.5.13}$$

then the conditional

$$\text{MSE}(p'\beta_{\epsilon_2}) \leq \text{MSE}(p'b). \tag{8.5.14}$$

Proof. If (8.5.13) and

$$(\beta - \theta)'(UU'F_2UU' + \sigma^2(X'X)^+)^+(\beta - \theta) \leq 1 \tag{8.5.15}$$

hold true (8.5.14) follows. Since $F_1 \leq F_2$, (8.5.13) implies that the inequality in (8.5.15) holds true. ∎

Since $F_1 \leq F_{\epsilon_2}$ the ellipsoid where $p'\beta_{\epsilon_1}$ has a smaller MSE than LS contains the ellipsoid where $p'\beta\epsilon_2$ has a smaller MSE than LS.

Example 8.5.1.   Mixture of BE and LS. Suppose there are two BE. The first one is derived from a prior with known mean θ_1 and known variance F_1. The second one is derived from a prior with unknown mean θ_2 and known variance F_2. The solution to the optimization problem is (for estimable parametric functions) from Rao(1973).

$$p'\hat{\beta}_\epsilon = \epsilon p'\hat{\beta}_1 + (1 - \epsilon)p'b,$$

i.e., a mixture of a linear BE and a LS estimator.

Example 8.5.2. The James-Stein Estimator as a Mixture of BE and LS. Let $\theta_1 = 0$ and $F_1 = (\sigma^2/k)(X'X)^+$ in Example 8.5.1 above. Then

$$p'\hat{\beta}_\epsilon = \epsilon p'\hat{\beta}_c + (1 - \epsilon)p'b$$

with

$$p'\beta_c = \frac{1}{1 + k}p'b = (1 - \frac{k}{1 + k})p'b.$$

When k is unknown, estimate $k/(1 + k)$ by $(\hat{\sigma}^2(s - 2))/(b'X'Xb)$. Then

$$p'\hat{\beta}_\epsilon = \epsilon(1 - \frac{\hat{\sigma}^2(s - 2)}{b'X'Xb})p'b + (1 - \epsilon)p'b$$

is the JS estimator.

Exercise 8.5.1. Show that if $s > 2$ in Example 8.5.2, there is an ϵ where $0 < \epsilon < 1$ and $\mathrm{MSE}(p'\hat{\beta}_\epsilon) \leq \mathrm{MSE}(p'b)$.

Exercise 8.5.2. Let $0 \leq \epsilon \leq 1$. Minimize

$$L =(Y - X\beta)'(Y - X\beta)\frac{1}{\sigma^2} + \frac{\epsilon}{\tau_1^2}(r_1 - R_1\beta)'(r_1 - R_1\beta)$$

$$+ \frac{1 - \epsilon}{\tau_2^2}(r_2 - R_2\beta)'(r_2 - R_2\beta)$$

to obtain the optimum estimator given the augmented model

$$
\begin{aligned}
Y &= X\beta + \epsilon, & E(\epsilon) &= 0, & D(\epsilon) &= \sigma^2 I, \\
r_1 &= R_1\beta + \phi_1, & E(\phi_1) &= 0, & D(\phi_1) &= \tau_1^2 I, \\
r_2 &= R_2\beta + \phi_2, & E(\phi_2) &= 0, & D(\phi_2) &= \tau_2^2 I.
\end{aligned}
$$

Find the mean and dispersion of the prior or the corresponding BE.

Exercise 8.5.3. Obtain the estimator in Exercise 8.5.2 above as a solution to the optimization problem for the generalized ridge regression estimator.

8.6 Summary

The MSE of the BE derived from incorrect prior information averaged over the correct prior information was obtained. Conditions for this MSE to be less than that of the LS estimator were given. The corresponding conditions were obtained for a minimax estimator where the MSE was maximized on an incorrect ellipsoid. Similar results were obtained for the mixed estimator given incorrect prior information.

The situation where the correct prior might not be known was considered using a mixture of priors. The MSE was compared with the least square estimator and with the case where the correct prior was known. The use of an estimator which took two different forms with certain probabilities was considered. Its MSE was greater than that used for an estimator derived from a mixed prior.

PART IV
APPLICATIONS

Chapter IX

The Kalman Filter

9.0 Introduction

9.0.1 Historical Perspective

The Kalman Filter has found successful application in many diverse areas. These include processing signals in aerospace tracking, underwater sonar, quality control, short term forecasting, analysis of life lengths from dose response experiments, determining a ship's position and many others.

During the 1960s many papers were published on the theory and applications of the Kalman Filter in journals whose primary readership were control engineers. The earliest papers on the subject were those of Kalman (1960) and Kalman and Bucy (1961).

More recent work has demonstrated the usefulness of the Kalman Filter to statisticians. Sorenson (1970) explains the relationship between the least square estimators and the Kalman Filter. Harrison and Stevens (1967) explains how the Kalman Filter may be viewed as a problem in Bayesian inference using some well known Box-Jenkins time series models as examples.

Many of the ideas to be developed in this chapter were motivated by Meinhold and Singpurwalla (1983). They show how the Kalman

Filter can be understood by statisticians by using a Bayesian formulation and some well known results from multivariate statistics. Their paper also includes illustrative examples from aerospace tracking and quality control.

A nice numerical example applying the Kalman Filter to the motion of a ship is considered by Lewis (1986), pp. 71–74.

Diderrich (1985) discusses the relationship between the Kalman Filter and the mixed estimator. He also shows how the ridge regression type estimators for a time varying linear model may be thought of as a "half Kalman Filter."

The problem of robustness of the Kalman Filter with respect to the initial prior assumptions is addressed in Soong (1965), Heffes (1966) and Nishimura (1966).

9.0.2 Goals of the Chapter

The discrete Kalman Filter is used to make inferences about a time varying linear model, given a dynamic linear stochastic relationship between the parameters at t and $t+1$. Given initial prior information at $t = 0$ the BE is calculated at $t = 1$. The BE, its prediction variance and the dynamic relationship between β_{t+1} and β_t is then used to formulate updated prior information. This process is then repeated.

This chapter will demonstrate how the concepts, methods and ideas described in Chapters III–VIII apply to the Kalman Filter. To do this it will:

1. give a formulation of the Kalman Filter using recursive Bayes estimation (see Section 9.1);
2. show how the Kalman Filter may be viewed as a recursive least squares estimator or a mixed estimator (see Section 9.2);
3. formulate the Kalman Filter as a minimax estimator (see Section 9.3);
4. explain the relationship between the Kalman Filter and the generalized ridge regression estimator (see Section 9.4);
5. study the properties of the MSE (see Section 9.5);
6. consider the robustness properties (see Section 9.6);

7. give examples of applications to quality control, aerospace tracking and ship movements (see Section 9.7);

8. study the relationship between the Kalman Filter and ridge type estimators (see Section 9.8). In this connection the James-Stein estimator will be presented as an empirical Bayes estimator. When knowledge about the prior distribution is either incomplete or not available sample estimates may be substituted for functions of the prior parameters in the BE. The estimators thus obtained are called empirical Bayes estimators.

Only the discrete Kalman Filter will be considered. Both the full and the non-full rank case will be considered.

9.1 The Kalman Filter as a Bayes Estimator

9.1.1 Overview of the Iterative Procedure

The Kalman Filter is an inference procedure that consists of:

1. a linear model defined at discrete times $t = 1, 2, 3 \ldots$;

2. a stochastic linear relation between the parameters at $t + 1$ and at t.

The following steps are followed:

3. Given this initial prior mean and variance using the linear relationship in 2 above the linear BE is obtained when $t = 1$.

4. From the relationship in 2 above the updated dispersion is obtained.

5. The linear BE and the updated dispersion in 4 serve as a "new" prior mean and dispersion.

6. Using the new prior mean and dispersion the linear BE is obtained again.

7. The process is repeated.

9.1.2 Comparison with Conventional Linear Model

Observe that the Kalman Filter differs from the conventional linear model because the regression coefficients are not constant; they change with time. The estimators at time t serve as prior information for the estimation of parameters at time $t + 1$. Given estimates of the parameters at time t the linear BE is obtained at time $t + 1$.

In light of the above discussion the mathematical presentation will now be given.

9.1.3 The Mathematical Formulation

1. For discrete time points $t = 0, 1, 2, \ldots$ consider a linear model

$$Y_t = X_t \beta_t + \epsilon_t, \tag{9.1.1}$$

where X_t is a fixed $n \times m$ matrix of rank $s_t \leq m$, β_t is an m dimensional random vector of parameters, ϵ_t is an n dimensional error vector and Y_t is an n dimensional vector of observations. The error vector ϵ_t satisfies

$$E(\epsilon_t | \beta_t) = 0 \quad \text{and} \quad D(\epsilon_t | \beta_t) = \sigma_t^2 I. \tag{9.1.2}$$

2. Notice that the β_t are random variables. The dynamic feature, i.e., the stochastic linear relationship between β_t and β_{t-1} is

$$\beta_t = M_t \beta_{t-1} + \eta_t, \tag{9.1.3}$$

where M_t is an $m \times m$ matrix. The η_t is an m dimensional error vector satisfying

$$E(\eta_t | \beta_t) = 0 \quad \text{and} \quad D(\eta_t | \beta_t) = W_t. \tag{9.1.4}$$

3. For $t = 0$ let the prior assumptions be

$$E(\beta_0) = \theta \quad \text{and} \quad D(\beta_0) = F_0. \tag{9.1.5}$$

4. From (9.1.3) notice that

$$E(\beta_1 | Y_0) = M_1 \theta \tag{9.1.6}$$

and

$$D(\beta_1|Y_0) = M_1 F_0 M_1' + W_1. \tag{9.1.7}$$

5. Let

$$\theta_1 = M_1\theta \quad \text{and} \quad F_1 = M_1 F_0 M_1' + W_1. \tag{9.1.8}$$

This is the "new" prior mean and dispersion.

6. The prior assumptions in (9.1.6) and (9.1.7) above are used to obtain the linear BE by the method of Chapter III. Thus,

$$p'\hat{\beta}_1 = p'M_1\theta + p'F_1 X_1'(X_1 F_1 X_1' + \sigma_1^2 I)^{-1}(Y_1 - X_1 M_1\theta). \tag{9.1.9}$$

7. The BE $p'\hat{\beta}_2$ is obtained for the prior assumptions

$$\theta_2 = M_2\hat{\beta}_1 \quad \text{and} \quad F_2 = M_2\Sigma_1 M_2' + W_2, \tag{9.1.10}$$

where

$$\Sigma_1 = E(\hat{\beta}_1 - \beta)(\hat{\beta}_1 - \beta)'$$

$$= F_1 - F_1 X_1'(\sigma_1^2 I + X_1 F_1 X_1')^{-1} X_1 F_1. \tag{9.1.11}$$

8. Once $p'\hat{\beta}_{t-1}$ is obtained $p'\hat{\beta}_t$ is the BE for the prior assumptions with

$$E(\beta_t) = M_t\hat{\beta}_{t-1} \quad \text{and} \quad D(\beta_t) = M_t\Sigma_{t-1} M_t' + W_t, \tag{9.1.12}$$

where

$$\Sigma_{t-1} = F_{t-1} - F_{t-1} X_{t-1}'(\sigma_t^2 I + X_{t-1} F_{t-1} X_{t-1}')^{-1} X_{t-1} F_{t-1}. \tag{9.1.13}$$

The BE takes the form

$$p'\hat{\beta}_t = p'\hat{\beta}_{t-1} + p'F_t X_t'(X_t F_t X_t' + \sigma_t^2 I)^{-1}(Y_t - X_t M_t\hat{\beta}_{t-1}). \tag{9.1.14}$$

The above derivations show how the Kalman Filter consists of an iterative BE where each iteration provides the prior information for the next step.

In Chapter III it was pointed out that the linear BE was the BE derived by Bayes Theorem if the population and the prior distribution both were normal. The estimators for the Kalman Filter derived by iterative application of Bayes Theorem would be the same as those derived above if the initial prior distribution and population are normal.

The matrix

$$P_t = F_t X_t'(X_t F_t X_t' + \sigma_t^2 I)^{-1} \qquad (9.1.15)$$

is called the Kalman gain.

Other authors (see for example Brown (1983)) obtain P_t as the matrix that minimizes the matrix

$$D_t = E(\beta_t - \beta_t)(\beta_t - \beta_t)'. \qquad (9.1.16)$$

The resulting estimators are the same as those obtained in (9.1.14) above since the optimum matrix P_t is that given in (9.1.15) above.

Example 9.1.1. Illustration of Iterative Procedure for Obtaining Estimates in the Kalman Filter. Let

$$Y_t = \mu_t + \epsilon_t,$$
$$\mu_t = \mu_{t-1} + \eta_t.$$

Assume that if $t = 0$ the prior distribution of μ has mean 0 and variance 4. Also assume

$$E(\epsilon_t|\mu_t) = 0, \quad D(\epsilon_t|\mu_t) = 1,$$
$$E(\eta_t|\mu_t) = 0, \quad D(\eta_t|\mu_t) = 2.$$

It is observed that

$$Y_1' = [1.1 \quad 1.2 \quad 1.6 \quad 1.4 \quad 1.8],$$

$$Y_2' = [2.0 \quad 2.7 \quad 2.4 \quad 2.3 \quad 2.2].$$

The parameters μ_1 and μ_2 are obtained by following the steps in (9.1.1) to (9.1.14). Since

$$E(\mu_0) = 0, \quad D(\mu_0) = 4,$$

from (9.1.5)

$$E(\mu_1|Y_0) = 0,$$
$$D(\mu_1|Y_0) = 4 + 2 = 6.$$

Now, from (9.1.8),

$$\theta_1 = 0 \quad \text{and} \quad F_1 = 6$$

is the new prior mean and dispersion. From (9.1.9) the estimate of μ_1 is

$$\hat{\mu}_1 = 4[1_5'][1_5 \cdot 4 \cdot 1_5' + I]^{-1} \begin{bmatrix} 1.1 \\ 1.2 \\ 1.6 \\ 1.4 \\ 1.8 \end{bmatrix}$$

$$= 4 \cdot 1_5'[4J + I]^{-1} \begin{bmatrix} 1.1 \\ 1.2 \\ 1.6 \\ 1.4 \\ 1.8 \end{bmatrix}$$

$$= 4 \cdot 1_5' \begin{bmatrix} 1.1 & -1.352 \\ 1.2 & -1.352 \\ 1.6 & -1.352 \\ 1.4 & -1.352 \\ 1.8 & -1.352 \end{bmatrix}$$

$$= 4 \cdot 1_5' \begin{bmatrix} -.252 \\ -.152 \\ .248 \\ .048 \\ .448 \end{bmatrix} = 1.36.$$

To obtain $\hat{\mu}_2$ observe that

$$\Sigma_1 = 6 - 6 \cdot 1'_5 (I + 1_5 \cdot 6 \cdot 1'_5)^{-1} 1_5 \cdot 6$$

$$= 6 - 61'_5 [I - \frac{6}{31} J] 1_5 \cdot 6 = \frac{6}{31}.$$

From (9.1.10),

$$\theta_2 = 0, \quad F_2 = \frac{6}{31} + 2 = \frac{68}{31},$$

and

$$\hat{\mu}_2 = \frac{68}{31} 1'_5 [1_5 \cdot \frac{68}{31} \cdot 1'_5 + I]^{-1} \begin{bmatrix} 2.0 \\ 2.7 \\ 2.4 \\ 2.3 \\ 2.2 \end{bmatrix}$$

$$= \frac{68}{31} \cdot 1'_5 [I - \frac{68}{371} J] \begin{bmatrix} 2.0 \\ 2.7 \\ 2.4 \\ 2.3 \\ 2.2 \end{bmatrix} = 2.126.$$

Exercise 9.1.1. Find $\hat{\mu}_3$ for Example 9.1.1 where

$$Y'_3 = [3.5 \quad 3.7 \quad 3.8 \quad 3.2 \quad 3.4].$$

Exercise 9.1.2. Repeat the work of Exercise 9.1.1 and Example 9.1.1 if $\theta' = [1 \quad -1 \quad 2 \quad -2 \quad 0]$ and $F_0 = 2$.

9.2 The Kalman Filter as a Recursive Least Square Estimator and the Connection with the Mixed Estimator

The Kalman Filter can be obtained recursively as a weighted least square estimator by using updating equations and the methods of Section 3.3. Recall in Section 3.3 that for the conventional regression model the mixed estimators were obtained as special weighted least square estimators.

In this section the Kalman Filter will be obtained as a weighted least square estimator. Both the full and the non-full rank case will be considered. For the full rank case a general M_t and X_t will be considered. To avoid certain difficulties the case of a constant X matrix and $M_t = I$ will be considered for the non-full rank case.

9.2.1 The Full Rank Case

Let H_t be a positive definite matrix. The following iterative process is used. The expression

$$L_1 = (\beta_1 - M_1\theta)'\frac{H_1}{\sigma_1^2}(\beta_1 - M_1\theta) + (Y_1 - X_1\beta_1)'(Y_1 - X_1\beta_1)\frac{1}{\sigma_1^2} \quad (9.2.1)$$

is minimized to obtain

$$p'\hat{\beta}_1 = p'(X_1'X_1 + H_1)^{-1}(X_1'Y_1 + H_1M_1\theta), \quad (9.2.2a)$$

where

$$\sigma_1^2 H_1^{-1} = M_1 H_0^{-1} M_1 \sigma_0^2 + W_1. \quad (9.2.2b)$$

Then

$$L_2 = (\beta_2 - M_2\hat{\beta}_1)'\frac{H_2}{\sigma_2^2}(\beta_2 - M_2\hat{\beta}_1) + (Y_2 - X_2\beta_2)'(Y_2 - X_2\beta_2)\frac{1}{\sigma_2^2}, \quad (9.2.3)$$

where

$$\sigma_2^2 H_2^{-1} = M_2\Sigma_1 M_2 + W_2, \quad (9.2.4)$$

is minimized to obtain

$$p'\hat{\beta}_2 = p'(X_2'X_2 + H_2)^{-1}(X_2'Y_2 + H_2M_2\hat{\beta}_1). \tag{9.2.5}$$

At the tth step

$$L_t = (\beta_t - M_t\hat{\beta}_{t-1})'\frac{H_t}{\sigma_t^2}(\beta_t - M_t\hat{\beta}_{t-1}) + (Y_t - X_t\beta_t)'(Y_t - X_t\beta_t)\frac{1}{\sigma_t^2} \tag{9.2.6}$$

is minimized to obtain

$$p'\hat{\beta}_t = p'(X_t'X_t + H_t)^{-1}(X_tY_t + H_tM_t\hat{\beta}_{t-1}). \tag{9.2.7}$$

When $H_t = \sigma_t^2 F_t^{-1}$ the estimator in (9.2.7) is equivalent to that of (9.1.14). The estimators (9.2.7) are least square estimators with respect to the augmented model

$$\begin{bmatrix} Y_t \\ M_t\hat{\beta}_{t-1} \end{bmatrix} = \begin{bmatrix} X_t \\ I \end{bmatrix}\beta_t + \begin{bmatrix} \epsilon_t \\ -\eta_t \end{bmatrix}, \tag{9.2.8}$$

where

$$E(\epsilon_t) = 0, \qquad D(\epsilon_t) = \sigma_t^2 I \tag{9.2.9}$$

and

$$E(\eta_t) = 0, \qquad D(\eta_t) = H_t^{-1}. \tag{9.2.10}$$

A matrix R_t can be found with

$$H_t = \left(\frac{\sigma_t^2}{\tau_t^2}\right)R_t'R_t, \tag{9.2.11}$$

$$r_t = R_tM_t\hat{\beta}_{t-1} = R_t\beta_t + \eta_t,$$

i.e., the regression of $\hat{\beta}_{t-1}$ on β_t .

The mixed estimator may be obtained as a mixed estimator from the linear model

$$\begin{bmatrix} Y_t \\ M_t\beta_{t-1} \end{bmatrix} = \begin{bmatrix} X_t \\ I \end{bmatrix}\beta_t + \begin{bmatrix} \epsilon_t \\ \eta_t \end{bmatrix} \tag{9.2.12}$$

with $r_t = M_t \hat{\beta}_{t-1}$ and

$$
\begin{aligned}
E(\epsilon_t) &= 0, & D(\epsilon_t) &= \sigma_t^2 I, \\
E(\eta_t) &= 0, & D(\eta_t) &= \tau_t^2 I.
\end{aligned}
\tag{9.2.13}
$$

Its form is

$$
p'\hat{\beta}_t = p'(\tau_t^2 X_t' X_t + \sigma_t^2 R_t' R_t)^{-1}(\tau_t^2 X_t' Y_t + \sigma_t^2 R_t' r_t),
\tag{9.2.14}
$$

where

$$
\sigma_t^2 (R_t' R_t)^{-1} = M_t \Sigma_{t-1} M_t + W_t
\tag{9.2.15a}
$$

with

$$
\Sigma_{t-1} = \sigma_{t-1}^2 (\tau_{t-1}^2 X_{t-1}' X_{t-1} + \sigma_{t-1}^2 R_{t-1}' R_{t-1})^{-1}.
\tag{9.2.15b}
$$

Equations (9.2.2b) and (9.2.15a) are motivated by the fact that (9.2.14) is Bayes with respect to a prior with mean $(R_t' R_t)^{-1} R' r_t$ and dispersion $(R_t' R_t)^{-1} \sigma_t^2$. Equation (9.2.15b) is just the variance of $p' \hat{\beta}_{t-1}$.

An example will now illustrate the ideas above.

Example 9.2.1. Kalman Filter Estimates. Consider the model

$$
Y_t = [0 \quad 1] \begin{bmatrix} \beta_{1t} \\ \beta_{2t} \end{bmatrix} + \epsilon_t,
$$

where

$$
E(\epsilon_t) = 0, \qquad D(\epsilon_t) = 1.
$$

together with

$$
\begin{bmatrix} \beta_{1t+1} \\ \beta_{2t+1} \end{bmatrix} = \begin{bmatrix} \frac{1}{2} & 1 \\ 0 & -\frac{1}{2} \end{bmatrix} \begin{bmatrix} \beta_{1t} \\ \beta_{2t} \end{bmatrix} + \begin{bmatrix} w_{1t} \\ w_{2t} \end{bmatrix},
$$

where

$$
E[W_t] = 0 \quad \text{and} \quad D[W_t] = \begin{bmatrix} 8 & 4 \\ 4 & 2 \end{bmatrix}.
$$

Assume that

$$\begin{bmatrix} \beta_{10} \\ \beta_{20} \end{bmatrix} = \begin{bmatrix} \frac{1}{2} \\ \frac{1}{2} \end{bmatrix} \quad \text{and} \quad H_0 = \begin{bmatrix} 3 & 0 \\ 0 & 2 \end{bmatrix}.$$

Now $p'\hat{\beta}_1$ will be obtained.

Equation (9.2.2b) for this case is

$$H_1^{-1} = \begin{bmatrix} \frac{1}{2} & 1 \\ 0 & -\frac{1}{2} \end{bmatrix} \begin{bmatrix} \frac{1}{3} & 0 \\ 0 & \frac{1}{2} \end{bmatrix} \begin{bmatrix} \frac{1}{2} & 0 \\ 1 & -\frac{1}{2} \end{bmatrix} + \begin{bmatrix} 8 & 4 \\ 4 & 2 \end{bmatrix}$$

$$= \begin{bmatrix} \frac{103}{12} & \frac{15}{4} \\ \frac{15}{4} & \frac{17}{8} \end{bmatrix}.$$

Thus,

$$H_1 = \frac{1}{401} \begin{bmatrix} 204 & -90 \\ 90 & 824 \end{bmatrix}.$$

Then

$$\hat{\beta}_1 = \frac{401}{241800} \begin{bmatrix} 1225 & 90 \\ 90 & 204 \end{bmatrix} \left[\begin{bmatrix} 0 \\ Y_{11} \end{bmatrix} + \frac{1}{401} \begin{bmatrix} \frac{1}{2} & 1 \\ 0 & -\frac{1}{2} \end{bmatrix} \begin{bmatrix} \frac{1}{2} \\ \frac{1}{2} \end{bmatrix} \right]$$

$$= \begin{bmatrix} 20315 & .1493 \\ .1493 & .3383 \end{bmatrix} \begin{bmatrix} .00249 \\ Y_{11} - 100.25 \end{bmatrix}$$

$$= \begin{bmatrix} .00506 + .1493(Y_{11} - 100.25) \\ .000372 + .3383(Y_{11} - 100.25) \end{bmatrix}$$

$$= \begin{bmatrix} .1493Y_{11} - 14.962 \\ .3383Y_{11} - 33.914 \end{bmatrix}.$$

Exercise 9.2.1. Obtain $\hat{\beta}_2, \hat{\beta}_3$ in Example 9.2.1 above, in terms of Y_1 and Y_2.

9.2.2 The Non-Full Rank Case

Assume that $X_t = X$ (the X_t are constant matrices) and $M_t = I$. Thus consider the linear model

$$Y_t = X\beta_t + \epsilon_t \qquad (9.2.16a)$$

together with the system equation

$$\beta_t = \beta_{t-1} + \eta_t. \qquad (9.2.16b)$$

When X and/or F_t are of non-full rank, the derivation in (9.2.1)–(9.2.7) still works. The result corresponding to (9.2.7) is

$$p'\hat{\beta}_t = p'(X'X + H_t)^+(X'Y_t + H_t\hat{\beta}_{t-1}). \qquad (9.2.17)$$

The dispersions are related by

$$\sigma_t^2 H_t^+ = \Sigma_{t-1} + W_t \qquad (9.2.18)$$

with

$$\Sigma_{t-1} = \sigma_{t-1}^2(\tau_{t-1}^2 X'X + \sigma_{t-1}^2 H_{t-1})^+.$$

In general (9.2.17) is not a solution to the optimization problems of finding a recursive BE. Theorem 9.2.1 will be helpful in obtaining conditions for (9.2.17) to be a recursive BE.

Theorem 9.2.1. If H_0 and W_t are of the form $U A_0 U'$ and $U C_t U'$, respectively, where A_0 and C_t are positive definite matrices then H_t is of the form $U A_t U'$ with A_t PD.

Proof. The proof is done by induction on t. When $t = 1$

$$\sigma_1^2 H_1^+ = \sigma_0^2 H_0^+ + W_1 = \sigma_0^2 U(A_0^{-1})U' + U C_1 U'$$

$$= U(\sigma_0^2 A_0^{-1} + C_1)U' \qquad (9.2.19)$$

and

$$H_1 = \sigma_1^2 U (\sigma_0^2 A_0^{-1} + C_1)^{-1} U'.$$

When $t = k$ assume the result is true. Now

$$\sigma_k^2 H_{k+1}^+ = \Sigma_k + U C_{k+1} U'. \tag{9.2.20}$$

Now $H_k = U A_k U'$ by induction hypothesis. Thus,

$$\Sigma_k = \sigma_k^2 (\tau_{k-1}^2 X'X + \sigma_{k-1}^2 H_k)^+$$

$$= \sigma_k^2 (\tau_{k-1}^2 U \Lambda U' + \sigma_{k-1}^2 U A_k U')^+ \tag{9.2.21}$$

$$= \sigma_k^2 U (\tau_{k-1}^2 \Lambda + \sigma_{k-1}^2 A_k)^{-1} U'.$$

This completes the induction. ∎

Thus, if $H_0 = U A_0 U'$ where A_0 is PD (9.2.17) is a BE at each stage with respect to a prior with mean $H_t \beta_{t-1}$ and dispersion $H_t^+ \sigma_t^2$.

Likewise when $U' F_t U$ is positive definite for all t the BE (9.1.14) may be written in the form (9.2.17), where $H_t = (U U' F_t U U')^+$. Thus (9.1.14) is equivalent to a recursive least square estimator or a mixed estimator. The following theorem is the analogue of Theorem 9.2.1.

Theorem 9.2.2. If $U' F_0 U$ is positive definite then $U' F_t U'$ is positive definite.

Proof. Again use induction on t. For $t = 1$,

$$F_1 = F_0 + W_1. \tag{9.2.22}$$

W_1 is a NND matrix. Since $U' F_0 U$ is PD, $U' F_1 U$ is PD. Assume result holds for $t = k$,

$$\Sigma_k = F_k - F_k (U U F_k U U' + \sigma_k^2 (X'X)^+)^+ F_k. \tag{9.2.23}$$

Then

$$U'\Sigma_k U = U'F_k U - U'F_k(UU'F_k UU' + \sigma_k^2(X'X)^+)^+ F_k U$$

$$= U'F_k U - U'F_k U(U'F_k U + \sigma_k^2\Lambda^{-1})^{-1}U'F_k U \quad (9.2.24)$$

$$= U'F_k U[(U'F_k U)^{-1} - (U'F_k U + \sigma_k^2\Lambda^{-1})^{-1}]U'F_k U$$

$$= U'F_k U'P_k U'F_k U, \qquad (9.2.25)$$

where

$$P_k = (U'F_k U)^{-1} - (U'F_k U + \sigma_k^2\Lambda^{-1})^{-1}.$$

Since

$$U'F_k U < U'F_k U + \sigma^2\Lambda^{-1}, \qquad (9.2.26)$$

P_k and thus $U'\Sigma_k U$ is positive definite. But,

$$F_{k+1} = \Sigma_k + W_{k+1} \qquad (9.2.27)$$

so $U'F_{k+1}U$ is PD. This completes the induction. ∎

Thus, (9.1.14) the BE and (9.2.17) the ridge type estimator are equivalent when $U'F_0 U$ is positive definite.

Now let R_t be such that

$$H_t = \frac{\sigma_t^2}{\tau_t^2}(R_t'R_t).$$

The mixed estimator takes the form

$$p'\hat{\beta}_t = p'(X'X\tau_t^2 + R_t'R_t\sigma_t^2)^+(X'Y_t\tau_t^2 + R_t'r_t\sigma_t^2) \qquad (9.2.28)$$

for the augmented linear model

$$\begin{bmatrix} Y_t \\ r_t \end{bmatrix} = \begin{bmatrix} X \\ R_t \end{bmatrix} \beta_t + \begin{bmatrix} \epsilon_t \\ \eta_t \end{bmatrix}. \qquad (9.2.29)$$

Also

$$E(\epsilon_t) = 0, \quad D(\epsilon_t) = \sigma_t^2 I,$$
$$E(\eta_t) = 0, \quad D(\eta_t) = \tau_t^2 I. \tag{9.2.30}$$

Furthermore

$$\sigma_t^2 (R_t' R_t)^+ = (R_{t-1}' R_{t-1})^+ \sigma_{t-1}^2 + W_t. \tag{9.2.31}$$

When $R_0 = A_0 U'$ and $W_t = U C_t U'$, where A_0 and C_t are positive definite matrices, (9.2.28) is equivalent to the BE.

9.3 The Minimax Estimator

The minimax version of the Kalman filter consists of an iterative application of the procedure in Section 3.3. Let θ and F_0 be the initial values of an m dimensional vector and $m \times m$ PD matrix. Let

$$\Omega_1 = \{\beta : (\beta - M_1 \theta)(\beta - M_1 \theta)' \le G_1\}. \tag{9.3.1}$$

Following the method of Section 3.4:

1. A linear estimator of the form

$$p'\hat{\beta}_1 = p'\theta + L_1'(Y_1 - X\theta) \tag{9.3.2}$$

is considered.
2. The maximum risk of (9.3.2) on ellipsoid (9.3.1) is obtained in terms of L_1.
3. The L_1 that minimizes the value of the expression obtained in 2 is found. Now letting

$$\theta_1 = M_1 \theta, \quad F_1 = M_1 F_0 M_1 + W_1 \quad \text{and} \quad G_1^+ = F_1, \tag{9.3.3}$$

the optimum estimator is obtained. For

$$\theta_2 = M_2 \hat{\beta}_1 \quad \text{and} \quad F_2 = M_2 \Sigma_1 M_2 + W_2,$$

where, on Ω_1,

$$\Sigma_1 = \max E_\beta(\hat\beta_1 - \beta_1)(\hat\beta_1 - \beta_1)', \qquad (9.3.4)$$

the optimum estimator is obtained again. At the tth step let

$$\Omega_t = \{\beta_t : (\beta_t - M_t\hat\beta_{t-1})'(\beta_t - M_t\hat\beta_{t-1})' \le F_t\}, \qquad (9.3.5)$$

where

$$F_t = M_t\Sigma_{t-1}M_t + W_t$$

and on Ω_{t-1}

$$\Sigma_{t-1} = \max E_\beta(\hat\beta_{t-1} - \beta_{t-1})(\hat\beta_{t-1} - \beta_{t-1})'.$$

From (3.4.3) the linear estimator of the form

$$p'\hat\beta_t = p'\hat\beta_{t-1} + L_t'(Y_t - X_t\hat\beta_{t-1}) \qquad (9.3.6)$$

has risk

$$\begin{aligned}
p'E_\beta(\hat\beta_t - \beta_t)(\hat\beta_t - \beta_t)'p \\
=\sigma_t^2 L_t'L_t \\
+ (L_t'X_t' - p_t')(\beta_t - M_t\beta_{t-1})(\beta_t - M_t\beta_{t-1})' \\
\times (X_tL_t - p_t).
\end{aligned} \qquad (9.3.7)$$

Its maximum on Ω_t is

$$p'Rp = (L'X_t - p')F_t(X_tL - p) + \sigma^2 L_t'L_t. \qquad (9.3.8)$$

The resulting estimator is

$$p'\hat\beta_t = p'\hat\beta_{t-1} + p'F_tX_t(X_tF_tX_t' + \sigma_t^2 I)^{-1}(Y_t - X_t\hat\beta_{t-1}). \qquad (9.3.9)$$

This is the minimax version of the Kalman Filter.

Remark. When F_t is positive definite Ω_t is equivalent to

$$\Omega_t = \{\beta_t : (\beta_t - M_t\hat\beta_{t-1})'F_t^{-1}(\beta_t - M_t\hat\beta_{t-1}) \le 1\}. \qquad (9.3.10)$$

When the X matrices are constant and $U'F_0U$, hence $U'F_tU$, is positive definite the minimax estimator may be found on the ellipsoid

$$(\beta_t - M_t\hat{\beta}'_{t-1})(UU'F_tUU')^+(\beta_t - M_t\hat{\beta}_{t-1}) \leq 1. \qquad (9.3.11)$$

Example 9.3.1. Ellipsoids for Minimax Estimators. Given the model in (9.2.1)

$$\Omega_0 = \{\beta_0 : (\beta_0 - \theta)' \begin{bmatrix} 3 & 0 \\ 0 & 2 \end{bmatrix} (\beta_0 - \theta) \leq 1\}$$

and

$$\Omega_1 = \{\beta_1 : (\beta_1 - M_1\theta) \begin{bmatrix} \frac{204}{401} & -\frac{90}{401} \\ -\frac{90}{401} & \frac{824}{401} \end{bmatrix} (\beta_1 - M_1\theta) \leq 1.$$

Exercise 9.3.1. Show that $F_1^{-1} - F_0^{-1}$ is not NND. Also show that the U matrices in the SVD are different so the axis of the two ellipsoids are not the same. Does one ellipsoid contain the other?

Exercise 9.3.2. Find r_2 and r_3 for Example 9.2.1. Compare the form of the ellipsoids.

Exercise 9.3.3. Given the conditions of Theorem 9.2.1 do all r_t have the same axis? Justify your answer.

9.4 The Generalized Ridge Estimator

The generalized ridge estimator was obtained in Section 3.2. It was obtained by finding the point on an ellipsoid centered at the least square estimator that had the minimum distance from a given point in the parameter space. The equations of the Kalman Filter may be obtained by solving the optimization problem of Section 3.2 iteratively. The fixed point in the parameter space at the tth step is the estimate at $t - 1$. The weight for the weighted distance to be minimized is found from past estimates via the Kalman gain equation.

Initially the weighted distance of B_1 from a point θ in the parameter space

$$D_1 = (B_1 - M_1\theta)' H_1(B_1 - M_1\theta) \tag{9.4.1}$$

is minimized, subject to

$$(B_1 - b_1)' X_1' X_1(B_1 - b_1) = \Phi_0, \tag{9.4.2}$$

where

$$b_1 = (X_1'X_1)^+ X_1'Y_1.$$

Letting λ_1 be the scalar Lagrange multiplier, differentiating

$$\begin{aligned} L_1 =& (B_1 - M_1\theta)' H_1(B_1 - M_1\theta) \\ &+ \lambda_1[(B_1 - b_1)' X_1' X_1(B_1 - b_1) - \Phi_1] \end{aligned} \tag{9.4.3}$$

and equating the result to zero,

$$H_1(B_1 - M_1\theta) + \lambda_1 X_1' X_1(\beta_1 - b_1) = 0 \tag{9.4.4}$$

or

$$(H_1 + \lambda_1 X_1' X_1)B_1 = H_1\theta + \lambda_1 X_1' X_1 b_1 \tag{9.4.5}$$

is obtained. Thus,

$$\hat{\beta}_1 = (H_1 + \lambda_1 X_1' X_1)^+ [H_1\theta + \lambda_1 X_1' X_1 b_1]. \tag{9.4.6}$$

Now on

$$S_1 = [\beta_1 : (\beta_1 - M_1\theta_0)(\beta_1 - M_1\theta_0)' \leq H_1] \tag{9.4.7}$$

let

$$p'\Sigma_1 p = \text{Var}(p'\hat{\beta}_1) + \text{Max Bias}(p'\hat{\beta}_1). \qquad (9.4.8)$$

Now

$$H_2 = [M_2\Sigma_1^+ M_2 + W_1]^+. \qquad (9.4.9)$$

Minimizing

$$L_2 = (B_2 - M_1\beta_1)'H_2(B_2 - M_1\beta_1) \qquad (9.4.10)$$

subject to

$$(B_2 - b_2)'X_2'X_2(B_2 - b_2) = \Phi_2, \qquad (9.4.11)$$

by following the same procedure as above

$$B_2 = (H_2^+\lambda_2 + X_2'X_2)^+[H_2\hat{\beta}_1 + \lambda_2 X_2'X_2 b_2] \qquad (9.4.12)$$

is obtained. At the tth step

$$D_t = (B_t - M_t\hat{\beta}_{t-1})'H_t(B_t - M_t\hat{\beta}_{t-1}) \qquad (9.4.13)$$

is minimized subject to

$$(B_t - b_t)'X_t'X_t(B_t - b_t) = \Phi_0. \qquad (9.4.14)$$

After differentiating and solving the matrix equation

$$\begin{aligned} L_t =& (B_t - M_t\beta_{t-1})'H_t(B_t - M_t\beta_{t-1}) \\ &+ \lambda_t[(B_t - b_t)'X_t'X_t(B_t - b_1) - \Phi_0] \end{aligned} \qquad (9.4.15)$$

the ridge estimator

$$\hat{\beta}_t = [H_t + \lambda_t X_t'X_t]^+[H_t\hat{\beta}_{t-1} + \lambda_t X_t'X_t b_t] \qquad (9.4.16)$$

is obtained.

Now

$$H_t = (M_t \Sigma_{t-1} M_t + W_{t-1})^+, \qquad (9.4.17)$$

where on

$$S_{t-1} = \{\beta_{t-1} : (\beta_{t-1} - M_{t-1}\hat{\beta}_{t-2})(\beta_{t-1} - M_{t-1}\hat{\beta}_{t-2})' \le H_{t-1}\},$$
$$(9.4.18)$$

$$p'\Sigma_{t-1}p = \mathrm{Var}(p'\hat{\beta}_{t-1}) + \mathrm{Max\ Bias}(p'\hat{\beta}_{t-1}). \qquad (9.4.19)$$

Exercise 9.4.1. For Example 9.2.1 obtain $H_1, H_2, H_3, B_1, B_2, B_3$ and compare your results to those of the example.

9.5 The Average MSE

In Chapter VI it was shown that more precise prior information led to a BE with smaller average MSE. For the Kalman Filter in Theorem 9.5.1 below it will be shown that more precise initial prior information gives an estimator with a smaller MSE for each value of t.

Theorem 9.5.1. *Consider two Kalman filters where the initial prior information is of the form*

$$\begin{aligned} E(\beta_0) &= \theta_{01}, \quad D(\beta_0) = F_{01}, \\ E(\beta_0) &= \theta_{02}, \quad D(\beta_0) = F_{02}. \end{aligned} \qquad (9.5.1)$$

Let $p'\hat{\beta}_{t1}$ and $p'\hat{\beta}_{t2}$ be the estimators derived at that t'th stage. If $F_{01} \le F_{02}$ then

$$\mathrm{MSE}(p'\hat{\beta}_{t1}) \le \mathrm{MSE}(p'\hat{\beta}_{t2}). \qquad (9.5.2)$$

Proof. The proof uses induction and Theorem 6.5.1. By Theorem 6.5.1, the result is true when $t = 1$. Assume it holds true for $t = k$ i.e.,

$$\Sigma_{k,1} \le \Sigma_{k,2}, \tag{9.5.3}$$

where $\Sigma_{k,i}$ $i = 1, 2$ is the MSE of $p'\hat{\beta}_{ki}$. Now

$$F_{k+1,1} = M_{k+1}\Sigma_{k1}M'_{k+1} + W_k \le M_{k+1}\Sigma_{k2}M'_{k+1} + W_k = F_{k+1,2}. \tag{9.5.4}$$

For the BE at stage $k + 1$, from Theorem 6.5.1,

$$\Sigma_{k+1,1} \le \Sigma_{k+1,2}. \ \blacksquare \tag{9.5.5}$$

Example 9.5.1. Comparison of Efficiencies of Two Kalman Filters. Consider the setup in (9.2.1) with

$$F_{01} = \begin{bmatrix} 1 & 0 \\ 0 & 1 \end{bmatrix} \quad \text{and} \quad F_{02} = \begin{bmatrix} 3 & 0 \\ 0 & 2 \end{bmatrix}.$$

Clearly $F_{02} - F_{01}$ is PD. Now, from (9.1.8),

$$F_{11} = \begin{bmatrix} \frac{1}{2} & 1 \\ 0 & -\frac{1}{2} \end{bmatrix} \begin{bmatrix} 1 & 0 \\ 0 & 1 \end{bmatrix} \begin{bmatrix} \frac{1}{2} & 0 \\ 1 & -\frac{1}{2} \end{bmatrix} + \begin{bmatrix} 8 & 4 \\ 4 & 2 \end{bmatrix}$$

$$= \begin{bmatrix} \frac{35}{4} & \frac{7}{2} \\ \frac{7}{2} & \frac{9}{4} \end{bmatrix}.$$

From (9.1.11)

$$\Sigma_{11} = \begin{bmatrix} \frac{35}{4} & \frac{7}{2} \\ \frac{7}{2} & \frac{9}{4} \end{bmatrix}$$

$$- \begin{bmatrix} \frac{35}{4} & \frac{7}{2} \\ \frac{7}{2} & \frac{9}{4} \end{bmatrix} \begin{bmatrix} 0 \\ 1 \end{bmatrix} \left[1 + \begin{bmatrix} 0 & 1 \end{bmatrix} \begin{bmatrix} \frac{35}{4} & \frac{7}{2} \\ \frac{7}{2} & \frac{9}{4} \end{bmatrix} \begin{bmatrix} 0 \\ 1 \end{bmatrix} \right]^{-1}$$

$$\times \begin{bmatrix} 0 & 1 \end{bmatrix} \begin{bmatrix} \frac{35}{4} & \frac{7}{2} \\ \frac{7}{2} & \frac{9}{4} \end{bmatrix}$$

$$= \begin{bmatrix} \frac{259}{52} & \frac{28}{26} \\ \frac{28}{26} & \frac{36}{52} \end{bmatrix}.$$

From (9.1.12)

$$F_{21} = \begin{bmatrix} \frac{1}{2} & 1 \\ 0 & -\frac{1}{2} \end{bmatrix} \begin{bmatrix} \frac{259}{52} & \frac{28}{26} \\ \frac{28}{26} & \frac{36}{52} \end{bmatrix} \begin{bmatrix} \frac{1}{2} & 0 \\ 1 & -\frac{1}{2} \end{bmatrix} + \begin{bmatrix} 8 & 4 \\ 4 & 2 \end{bmatrix}$$

$$= \begin{bmatrix} \frac{1459}{104} & \frac{352}{104} \\ \frac{352}{104} & \frac{226}{104} \end{bmatrix},$$

$$F_{12} = \begin{bmatrix} \frac{1}{2} & 1 \\ 0 & -\frac{1}{2} \end{bmatrix} \begin{bmatrix} 3 & 0 \\ 0 & 2 \end{bmatrix} \begin{bmatrix} \frac{1}{2} & 0 \\ 1 & -\frac{1}{2} \end{bmatrix} + \begin{bmatrix} 8 & 4 \\ 4 & 2 \end{bmatrix}$$

$$= \begin{bmatrix} \frac{43}{4} & 3 \\ 3 & \frac{5}{2} \end{bmatrix},$$

$$\Sigma_{12} = \begin{bmatrix} \frac{43}{4} & 3 \\ 3 & \frac{5}{2} \end{bmatrix}$$

$$- \begin{bmatrix} \frac{43}{4} & 3 \\ 3 & \frac{5}{2} \end{bmatrix} \begin{bmatrix} 0 \\ 1 \end{bmatrix} \left[1 + \begin{bmatrix} 0 & 1 \end{bmatrix} \begin{bmatrix} \frac{43}{4} & 3 \\ 3 & \frac{5}{2} \end{bmatrix} \begin{bmatrix} 0 \\ 1 \end{bmatrix} \right]^{-1}$$

$$\times \begin{bmatrix} 0 & 1 \end{bmatrix} \begin{bmatrix} \frac{43}{4} & 3 \\ 3 & \frac{5}{2} \end{bmatrix}$$

$$= \begin{bmatrix} \frac{229}{28} & \frac{6}{7} \\ \\ \frac{6}{7} & \frac{10}{14} \end{bmatrix},$$

and

$$F_{22} = \begin{bmatrix} \frac{229}{28} & \frac{6}{7} \\ \\ \frac{6}{7} & \frac{10}{14} \end{bmatrix} + \begin{bmatrix} 8 & 4 \\ \\ 4 & 2 \end{bmatrix} = \begin{bmatrix} \frac{453}{28} & \frac{34}{7} \\ \\ \frac{34}{7} & \frac{38}{14} \end{bmatrix}.$$

Now

$$F_{22} - F_{21} = \begin{bmatrix} 2 & -\frac{1}{2} \\ \\ -\frac{1}{2} & \frac{1}{4} \end{bmatrix}, \quad \text{a PD matrix.}$$

Also

$$\Sigma_{12} - \Sigma_{11} = \begin{bmatrix} 3.1978 & -.21978 \\ -.21978 & .021978 \end{bmatrix} \text{ is a PD matrix.}$$

Exercise 9.5.1. Find Σ_{21}, Σ_{22}, F_{32}, and F_{31}. Show that $\Sigma_{22} - \Sigma_{21}$ and $F_{32} - F_{31}$ are PD.

Exercise 9.5.2. Are $F_{21} - F_{11}$ and $\Sigma_{21} - \Sigma_{11}$ a definite matrix? What implications might this have?

Exercise 9.5.3. For Example 9.2.1, what is the MSE($\hat{\beta}_{2t}$) when t=1, 2, or 3 where H_0 is as given in that example?

9.6 The MSE for Incorrect Initial Prior Assumptions

In Chapter VIII the robustness of the BE where the prior mean
and dispersion were misspecified was considered. The MSE of a BE
derived from incorrect prior information was obtained averaging over
the correct prior information. Conditions were derived for this MSE
to be smaller than that of the LS estimator.

Similar results will now be obtained for the Kalman Filter when
the initial prior assumptions are incorrect.

Suppose that the prior assumptions

$$E(\beta_0) = \theta \quad \text{and} \quad D(\beta_0) = F_0 \tag{9.6.1}$$

are incorrect. Then the prior assumptions used at stage $t + 1$ will
also be incorrect. The MSE of the estimator derived for incorrect
prior assumptions will be obtained averaging over the correct prior
assumptions.

Assume that the correct prior assumptions are

$$E(\beta_0) = \eta_0 \quad \text{and} \quad D(\beta_0) = H_0. \tag{9.6.2}$$

At stage 1 the correct prior assumptions should then be

$$E(\beta_1|Y_0) = M_1\theta = \eta_1 \tag{9.6.3}$$

and

$$D(\beta_1|Y_0) = M_1 H_0 M_1 + W_1 = H_1. \tag{9.6.4}$$

Thus,

$$p'\hat{\beta}_1 = p'\eta_1 + p'H_1 X_1 (X_1 H_1 X_1' + \sigma^2 I)^{-1}(Y_1 - X_1\eta_1). \tag{9.6.5}$$

At stage t,

$$\eta_t^c = M_t\hat{\beta}_{t-1} \quad \text{and} \quad H_t = M_t\Sigma_{t-1}M_t + W_t, \tag{9.6.6}$$

where $\hat{\beta}c_{t-1}$ is the estimator obtained for the correct prior assumptions.

From (8.1.4) the MSE of $p'\hat{\beta}_t$ derived from the incorrect prior assumptions averaged over the correct prior assumptions is

$$
\begin{aligned}
\text{MSE} =p'\Gamma_{1t}^{+}[&\sigma_t^2 X_t'X_t + (UUF_tU_tU_t')^{+} \\
&\times [H_t + M_t(\hat{\beta}_{t-1} - \hat{\beta}_{t-1}^c)(\hat{\beta}_{t-1} - \hat{\beta}_{t-1}^c)'M_t] \qquad (9.6.7) \\
&\times (U_tU_t'F_tU_tU_t')^{+}\sigma_t^4]\Gamma_{1t}^{+}p,
\end{aligned}
$$

where

$$
\Gamma_{1t} = X_t'X_t + \sigma_t^2(U_tU_t'F_tU_tU_t')^{+}. \qquad (9.6.8)
$$

Using (9.6.8) the analogues of Theorems 8.1.1 and 8.1.2 may be derived. These results are stated in Theorems 9.6.1 and 9.6.2 below.

Theorem 9.6.1. *If either of the following conditions hold true:*

(1) $H_t + (\beta_t - M_t\hat{\beta}c_{t-1})'(\beta_t - M_t\hat{\beta}c_{t-1}) \le F_t$ \qquad (9.6.9)

or

(2) $H_t \le F_t, \beta_t - M_t\hat{\beta}c_{t-1}$ *belongs to the range of* $F_t - H_t$ *and*

$$
(\beta_t - M_t\hat{\beta}c_{t-1})'(F_t - H_t)^{+}(\beta_t - M_t\hat{\beta}c_{t-1}) \le 1, \qquad (9.6.10)
$$

the MSE for the correct prior assumptions (9.6.2) is less than the MSE for the incorrect prior assumptions (9.6.1).

Theorem 9.6.2. *The MSE*

$$
\tilde{M}_3 \le \text{MSE}(p'\hat{\beta}_t)
$$

iff

(1) $2U_t'F_tU_t + \sigma_t^2\Lambda_t^{-1} - U_tH_tU_t' > 0,$ \qquad (9.6.11)

and

(2) β_t *lies in the ellipsoid* (9.6.9).

When $H_t = F_t$ the condition reduces to

$$
(\beta_t - M_t\hat{\beta}_{t-1})'[\sigma_t^2(X_t'X_t)^{+} + U_tU_t'F_tU_tU_t']^{+}(\beta_t - M_t\hat{\beta}_{t-1}) \le 1.
$$
$$
\tag{9.6.12}
$$

Exercise 9.6.1. Suppose that in Example 9.2.1 a prior where

$$
\begin{bmatrix} \beta_{10} \\ \beta_{20} \end{bmatrix} = \begin{bmatrix} \frac{1}{2} \\ \frac{1}{2} \end{bmatrix} \quad \text{and} \quad F_0 = \begin{bmatrix} \frac{1}{3} & 0 \\ 0 & \frac{1}{4} \end{bmatrix}
$$

is incorrect. The correct prior is

$$
\begin{bmatrix} \beta_{10} \\ \beta_{20} \end{bmatrix} = \begin{bmatrix} 0 \\ 1 \end{bmatrix} \quad \text{and} \quad F_0 = \begin{bmatrix} \frac{1}{4} & 0 \\ 0 & \frac{1}{3} \end{bmatrix}.
$$

What is the explicit form of the ellipsoids in Theorem 9.6.1 and 9.6.2 for t=0, 1, 2, or 3 ?

9.7 Applications

Three examples of applications will be described here. They include;

1. an illustration of the Kalman Filter in quality control;
2. how the Kalman Filter may be used for satellite tracking;
3. an example with numerical estimates from navigation.

The situations in Subsections 9.7.1 and 9.7.2 are discussed by Meinhold and Singpurwalla (1983).

9.7.1 Quality Control

Let Y_t be a simple transform of the number of defectives observed in a sample obtained at time t. Let θ_{1t} represent the true defective index of the process. Let θ_{2t} represent the drift of this process. The observation equation is

$$
Y_t = \theta_{1t} + v_t. \tag{9.7.1}
$$

The system equations are

$$
\theta_t = \begin{bmatrix} 0 & 1 \\ 0 & 1 \end{bmatrix} \begin{bmatrix} \theta_{1t-1} \\ \theta_{2t-1} \end{bmatrix} + \begin{bmatrix} 1 & 1 \\ 0 & 1 \end{bmatrix} \begin{bmatrix} w_{1t} \\ w_{2t} \end{bmatrix}. \tag{9.7.2}
$$

To simplify this example of the Kalman Filter, the above setup is considered without the drift parameter. This yields

$$Y_t = \theta_t + v_t \qquad (9.7.3a)$$

together with

$$\theta_t = \theta_{t-1} + w_t. \qquad (9.7.3b)$$

Here $X_t = 1$, $M_t = 1$, $\Sigma_0 = 1$, $V_t = 2$, $W_t = 1$, $F_t = 2$ and $\Sigma_t = 1$. Thus, the BE is

$$\hat{\theta}_t = \hat{\theta}_{t-1} + \frac{1}{2}(Y_t - \hat{\theta}_{t-1})$$

$$= \sum_{j=0}^{t-1} (\frac{1}{2})^{j+1} Y_{t-j} + (\frac{1}{2})^t \theta_0. \qquad (9.7.4)$$

9.7.2 Satellite Tracking

Let θ_t represent an unknown state of nature, e.g., the position and speed at time t with respect to a spherical coordinate system with origin at the center of the earth. Position and speed of the satellite cannot be measured directly. The measurements of the distance to the satellite and the accompanying angles of measurement are obtained from tracking stations around the earth. These are represented by Y_t. Principles of geometry are used to map Y_t into θ_t and are incorporated into the design matrix X_t. The ε_t represents the measurement error. Thus, the model is

$$Y_t = X_t\theta_t + \varepsilon_t. \qquad (9.7.5)$$

The M_t in the system equation prescribes the position and the speed change according to the laws of physics concerned with orbiting bodies. The η_t allows for deviations from these laws due to phenomena like the nonuniformity of the earth's magnetic field. Thus, the system equation is

$$\theta_t = M_t\theta_{t-1} + \eta_t. \qquad (9.7.6)$$

9.7.3 Navigation

This example is taken from Lewis (1986), pp.72–74. A ship moves east at 10 mph. Except for the effect of the wind gusts and wave actions the ship's velocity is assumed to be constant. Each hour an estimate of position d_k and velocity s_k is required.

The navigator assumes that the initial prior distribution of the position is normal with mean $d_0 = 0$ and variance $\sigma_0^2 = 2$. The initial velocity also has a normal prior distribution; its mean is 10 ($s_0 = 10$) and its variance $\sigma_0^2 = 3$. The objective is to find the Bayes estimator of position and velocity d_k and s_k.

The system dynamics and the measurement process must be modelled. During hour k, since the ship moves with velocity s_k mph its position changes according to

$$d_{k+1} = d_k + s_k. \tag{9.7.7}$$

Taking into account the unknown effects of wind and waves

$$s_{k+1} = s_k + w_k. \tag{9.7.8}$$

Assume w_k is normally distributed. Navigational fixes are taken at times $k = 1, 2, 3$. They determine position to within an error covariance of 2 but give no direct information about velocity.

Now let

$$\beta_k = \begin{bmatrix} d_k \\ s_k \end{bmatrix}. \tag{9.7.9}$$

The output equation is

$$y_k = [1 \ 0]\beta_k + v_k \quad \text{with} \quad v_k \sim N(0, 2) \tag{9.7.10}$$

because only the positions are observed and v_k is the error of measurement. The dynamic equation is

$$\beta_{k+1} = \begin{bmatrix} 1 & 1 \\ 0 & 1 \end{bmatrix} \beta_k + \begin{bmatrix} 0 \\ 1 \end{bmatrix} w_k, \tag{9.7.11}$$

where

$$w_k \sim N(0,1).$$

Observations at the first three time points give easterly positions of $y_1 = 9$, $y_2 = 19.5$ and $y_3 = 29$.

The estimates of β_1, β_2 and β_3 will be obtained. Observe that for the initial prior

$$\theta = \begin{bmatrix} 0 \\ 10 \end{bmatrix} \quad \text{and} \quad F_0 = \begin{bmatrix} 2 & 0 \\ 0 & 2 \end{bmatrix}.$$

Now

$$M = \begin{bmatrix} 1 & 1 \\ 0 & 1 \end{bmatrix}, \quad \theta_1 = M_1\theta = \begin{bmatrix} 1 & 0 \\ 1 & 0 \end{bmatrix}.$$

Thus,

$$F_1 = \begin{bmatrix} 1 & 1 \\ 0 & 1 \end{bmatrix} \begin{bmatrix} 2 & 0 \\ 0 & 3 \end{bmatrix} \begin{bmatrix} 1 & 0 \\ 1 & 1 \end{bmatrix} + \begin{bmatrix} 0 \\ 1 \end{bmatrix} [0 \quad 1] = \begin{bmatrix} 5 & 3 \\ 3 & 4 \end{bmatrix}$$

and

$$\Sigma_1 = \begin{bmatrix} 1.429 & 0.857 \\ 0.857 & 2.714 \end{bmatrix}.$$

Thus,

$$\hat{\beta}_1 = \begin{bmatrix} 9.286 \\ 9.571 \end{bmatrix}. \tag{9.7.12}$$

Now

$$\theta_2 = M_2\hat{\beta}_1 = \begin{bmatrix} 1 & 1 \\ 0 & 1 \end{bmatrix} \begin{bmatrix} 9.286 \\ 9.571 \end{bmatrix} = \begin{bmatrix} 18.857 \\ 9.571 \end{bmatrix}$$

and

$$F_2 = \begin{bmatrix} 1 & 1 \\ 0 & 1 \end{bmatrix} \begin{bmatrix} 1.429 & 0.857 \\ 0.857 & 2.714 \end{bmatrix} \begin{bmatrix} 1 & 0 \\ 1 & 1 \end{bmatrix} + [0 \quad 1] \begin{bmatrix} 0 \\ 1 \end{bmatrix}$$

$$= \begin{bmatrix} 5.857 & 3.571 \\ 3.571 & 3.714 \end{bmatrix} \cdot$$

Then

$$\Sigma_2 = \begin{bmatrix} 1.491 & 0.909 \\ 0.909 & 2.091 \end{bmatrix}$$

and

$$\hat{\beta}_2 = \begin{bmatrix} 19.336 \\ 9.864 \end{bmatrix} \cdot \qquad (9.7.13)$$

Repeating the process

$$\theta_3 = \begin{bmatrix} 29.2 \\ 9.864 \end{bmatrix},$$

$$F_3 = \begin{bmatrix} 5.4 & 3 \\ 3 & 3.091 \end{bmatrix}$$

and

$$\hat{\beta}_3 = \begin{bmatrix} 29.054 \\ 9.783 \end{bmatrix} \cdot \qquad (9.7.14)$$

This process may be continued.

Exercises 9.7.1–9.7.5 refer to the example above.

Exercise 9.7.1. Estimate β_4 and β_5 given $y_4 = 38.5$ and $y_5 = 47$.

Exercise 9.7.2. Redo the example given $y_1 = 5$, $y_2 = 10$, $y_3 = 15$.

Exercise 9.7.3. Let

$$H_0^1 = \begin{bmatrix} 1 & 0 \\ 0 & 1 \end{bmatrix}, \quad H_0^2 = \begin{bmatrix} 3 & 0 \\ 0 & 2 \end{bmatrix};$$

verify by direct computation that $\Sigma_{31} \leq \Sigma_{32}$.

Exercise 9.7.4. If the estimators in the example of this subsection are viewed as minimax estimators give the ellipsoids $\Omega_1, \Omega_2, \Omega_3$.

Exercise 9.7.5. In formulation generalized ridge estimators give the matrix H_1, H_2, H_3 and $\hat{\beta}_1, \hat{\beta}_2, \hat{\beta}_3$. Are the estimators the same as the BE and minimax estimators?

9.8 Recursive Ridge Regression

The ridge regression estimators in the context of the Kalman Filter will now be developed for the case of constant X matrices and $M_t = I$. Thus, consider the linear model

$$Y_t = X\beta_t + \varepsilon_t \tag{9.8.1}$$

with

$$E(\varepsilon_t|\beta_t) = 0 \quad \text{and} \quad D(\varepsilon_t|\beta_t) = \sigma_t^2 I. \tag{9.8.2}$$

The β_t and β_{t-1} are related by

$$\beta_t = \beta_{t-1} + \eta_t. \tag{9.8.3}$$

Recall that the ridge estimators were derived for prior distributions with dispersions of the form UDU' with D a diagonal matrix. The goal is to have updated priors with dispersions in this form. Thus, assume

$$E(\eta_t|\beta_t) = 0 \quad \text{and} \quad D(\eta_t|\beta_t) = U\Delta_t^{-1}U' \qquad (9.8.4)$$

with Δ_t a diagonal matrix. Also assume

$$E(\beta_0) = 0 \quad \text{and} \quad D(\beta_0) = UD_0^{-1}U'\sigma_0^2 = F_0, \qquad (9.8.5)$$

where D_0 is a positive definite matrix. Now

$$E(\beta_1|Y_0) = 0 \quad \text{and} \quad D(\beta_1|Y_0) = UD_0^{-1}U + U\Delta_1^{-1}U'. \qquad (9.8.6)$$

The BE

$$p'\beta_1 = p'F_1X(XF_1X' + \sigma_1^2I)^{-1}Y_1 = p'(X'X + \sigma_1^2UD_1U')^+X'Y_1. \qquad (9.8.7)$$

Observe from (9.8.6)

$$D_1^{-1} = D_0^{-1} + \Delta_1^{-1}. \qquad (9.8.8)$$

Also

$$D_t^{-1} = D_{t-1}^{-1} + U'\Sigma_{t-1}U. \qquad (9.8.9)$$

At stage t the ridge estimator is

$$p'\hat{\beta}_t = p'(X'X + \sigma_t^2UD_tU')^+X'Y_t. \qquad (9.8.10)$$

From (9.8.9) the ith diagonal element satisfies

$$\frac{1}{d_{ti}} = \frac{\sigma_{t-1}^2}{\lambda_i + d_{t-1i}} + \frac{1}{\delta_{ti}} = \frac{\sigma_{t-1}^2\delta_{ti} + \lambda_i + d_{t-1i}}{\delta_{ti}(\lambda_i + d_{t-1i})}. \qquad (9.8.11)$$

Then

$$d_{ti} = \frac{\delta_{ti}(\lambda_i + d_{t-1i})}{\sigma_{t-1}^2\delta_{ti} + \lambda_i + d_{t-1i}}. \qquad (9.8.12)$$

For constant δ_{ti} and σ^2 the steady state solution would be the positive roots of the equation

$$x^2 + (\sigma_{t-1}^2\delta_{ti} + \lambda_i - \delta_{ti})x - \delta_{ti}\lambda_i = 0 \qquad (9.8.13)$$

for each i. Conditions need to be determined for the sequence d_{ti} to converge to the steady state solution. A ridge regression estimator would be Bayes with respect to a prior whose dispersion is of the the form

$$F = UD^{-1}U'\sigma^2\delta,$$

where D is a diagonal matrix whose elements satisfy (9.8.13).

Recall that for the ordinary ridge estimator the diagonal matrix D had equal elements. If the initial prior is of the form UDU' where D has equal elements, generally the updated prior will not have this property. Thus the estimator obtained at stage t will be a ridge type estimator but not an ordinary ridge estimator.

Contraction estimators may be obtained by letting

$$W_t = \frac{\sigma_t^2}{\delta_t}(X'X)^+ \tag{9.8.14}$$

and

$$F_0 = \frac{\sigma^2}{k_0}(X'X)^+. \tag{9.8.15}$$

Then

$$\Sigma_t = \sigma_t^2(k_t(X'X) + (X'X))^+ = \frac{\sigma_t^2}{1+k_t}(X'X)^+. \tag{9.8.16}$$

Thus

$$F_t = \left[\frac{\sigma_t^2}{\delta_t} + \frac{\sigma_{t-1}^2}{k_{t-1}+1}\right](X'X)^+ = \frac{\sigma_t^2}{k_t}(X'X)^+ \tag{9.8.17}$$

and

$$\frac{\sigma_t^2}{\delta_t} + \frac{\sigma_{t-1}^2}{k_{t-1}+1} = \frac{\sigma_t^2}{k_t} \tag{9.8.18}$$

or

$$k_t = \frac{\sigma_t^2\delta_t(1+k_{t-1})}{\sigma_t^2 k_{t-1} + \sigma_{t-1}^2\delta_t + \sigma_t^2}. \tag{9.8.19}$$

The contraction estimator at stage t is

$$p'\hat{\beta}_{tc} = \frac{p'b_t}{1 + k_t}. \tag{9.8.20}$$

The James-Stein estimator at the tth stage is

$$p'\hat{\beta}_t = \left[1 - \frac{\hat{\sigma}_t^2 d}{b_t' X' X b_t}\right] p'b_t, \tag{9.8.21}$$

where

$$\hat{\sigma}_t^2 = \frac{1}{r(n-s)}[Y_t'Y_t - Y_t'Xb_t]. \tag{9.8.22}$$

Exercise 9.8.1. For the contraction estimator suppose $\sigma_t^2 = \sigma^2$ for all t and $\delta_t = 2$. Show that for the steady state, if it exists, $k_t = 1$.

Exercise 9.8.2. Write down the analogues of equations (9.8.15)–(9.8.19) for the generalized contraction estimator (4.5.13). What can you say about the steady state if it exists? Assume

$$W_t = \frac{\sigma_t^2}{\delta_t}(U\Lambda^{-1}K_t^{-1}U'),$$

where k_t is a positive definite diagonal matrix.

9.9 Summary

The Kalman Filter is an inference procedure consisting of a time varying linear model, a stochastic relationship and initial assumptions about the distribution. The Bayes estimator is obtained. Together with the stochastic relationship updated prior information is obtained and the BE is found using this updated prior information.

By assigning proper non-Bayesian roles to the mean vector and dispersion matrix and solving optimization problems it was shown how the Kalman Filter could be regarded as a mixed, recursive least square, generalized ridge or minimax estimator. Examples of applications were given and the connection with ridge type estimators was described.

Chapter X

Experimental Design Models

10.0 Introduction

Important special cases of the linear models considered in this book arise when the design matrix consists of ones and zeros only. The special cases include most of the standard Analysis of Variance (ANOVA) models.

Two basic types of Analysis of Variance (ANOVA) are the one and two way classification. In the one way classification only one factor is investigated. In the two way classification two factors are investigated.

In the one way ANOVA the mean effect of several treatments is compared to see if there is a significant difference between them. It is required that the experiment be performed in a random order so that the environment is as uniform as possible. The experimental design is thus called a completely randomized design.

A special case of the two way ANOVA is encountered when it is necessary to systematically control external sources of variation. This may be done by arranging the data in blocks where each block will contain one observation per treatment. This experimental design is called the randomized block design.

Another case of the two way ANOVA occurs when each of two factors have several levels. The experiment is done for all possible combinations of levels of the two factors; the ordering of the runs is random. There may be an interaction between the two factors. This is called a factorial experiment.

The main purpose of this chapter is to show how some of the ideas in Chapters III–VIII may be applied to experimental designs.

In linear statistical models the significance of the different factors represented by the independent variables are tested for significance by a technique called Analysis of Variance (ANOVA). In ANOVA the sum of the squares of the differences between the observed values and their means is broken up into sums of squares due to the different factors and experimental error. Significance tests are based on the ratio of the sum of squares due to the factors and the error sum of squares. The expressions for these sums of squares will prove to be useful in formulating approximate Bayes estimators. The form of the linear model for one way ANOVA, the least square estimators and a characterization of the estimable parametric functions is given in Section 10.1. The ANOVA technique is briefly reviewed.

Sometimes the treatments are fixed. At other times they are drawn from a large collection of treatments and the parameters are assumed to be random variables. Thus, there are two types of models: the fixed effects model with constant parameters and the random effects (variance component) model. Using the assumptions about the coefficients as random variables for prior information, Bayes estimators are derived in Section 10.2.

Often prior information is not available or complete. The function of the prior parameters may then be estimated by sample estimates. These sample estimates may then be substituted for the functions of the prior parameters in the BE. The resulting approximate BE are called empirical Bayes estimators EBE. The EBEs and their average MSE are also presented in Section 10.2. The computation of the MSE is outlined in an appendix.

The analogous results are obtained for the randomized block design in Sections 10.3 and 10.4. Using some results on patterned matrices (see Section 2.6) the form of the BE and the EBE is derived. The form of the MSE is presented.

The average MSE of the BE and the EBE are both smaller than that of the least square estimator. A measure of the efficiency of an EBE as compared with a BE is the Relative Savings Loss RSL (see

Efron and Morris, 1973). This measure is calculated for the EBE's in Sections 10.2 and 10.4.

10.1 The One Way ANOVA Model

10.1.1 The Model and the LS Estimates

Consider the linear model

$$Y = [1_a \otimes 1_n \quad I_a \otimes 1_n] \begin{bmatrix} \mu \\ \alpha \end{bmatrix} + \varepsilon. \qquad (10.1.1)$$

Let $N = na$. The vector $\mu = 1_N \cdot \mu$ where μ is scalar. The vector α is $\alpha' = (\alpha_1, \alpha_2, \cdots, \alpha_a)$. The vector Y is an N dimensional vector of observations and ε is an N dimensional error vector. The normal equations are (in matrix form)

$$\begin{bmatrix} an & na \\ n1_a & nI_a \end{bmatrix} \begin{bmatrix} \mu \\ \alpha \end{bmatrix} = \begin{bmatrix} 1'_a \otimes 1'_n \\ I_a \otimes 1'_n \end{bmatrix} Y \qquad (10.1.2)$$

or

$$an\mu + n \sum_{i=1}^{a} \alpha_i = Y_{..},$$

$$(10.1.3)$$

$$n\mu + n\alpha_i = Y_{i.}, \quad 1 \le i \le a,$$

where

$$Y_{..} = \sum_{i=1}^{a} \sum_{j=1}^{n} Y_{ij} \text{ and } Y_{i.} = \sum_{j=1}^{n} Y_{ij}.$$

Some authors, e.g. Montgomery (1984), solve these normal equations subject to the constraint

$$\sum_{i=1}^{a} \alpha_i = 0.$$

Thus,

$$\hat{\mu} = \overline{Y}_{..} \qquad\qquad (10.1.4)$$

and

$$\hat{\alpha}_i = \overline{Y}_{i.} - \overline{Y}_{..}, \quad 1 \le i \le a.$$

The estimable parametric functions $p'\beta$ have two very important properties.

1. They are independent of the choice of the generalized inverse.
2. They include the orthogonal contrasts (i.e., functions of the form $c'\alpha$ where

$$\sum_{i=1}^{a} c_i = 0).$$

Hypothesis of the form $c'\alpha = 0$ where the $c'\alpha$ are orthogonal contrasts are often tested in the solution of practical problems.

Recall that in general the form of the least square estimator is

$$\hat{\beta} = \begin{bmatrix} \hat{\mu} \\ \hat{\alpha} \end{bmatrix} = GX'Y. \qquad\qquad (10.1.5)$$

For the solutions in (10.1.4) the generalized inverse of $X'X$,

$$G = \begin{bmatrix} \frac{1}{na} & 0 \\ -\frac{1}{na}I_a & \frac{1}{n}I_a \end{bmatrix}. \qquad\qquad (10.1.6)$$

10.1.2 Characterization of the Estimable Parametric Functions

Recall that the SVD of

$$X'X = [U\ V] \begin{bmatrix} \Lambda & 0 \\ 0 & 0 \end{bmatrix} \begin{bmatrix} U' \\ V' \end{bmatrix}. \qquad\qquad (10.1.7)$$

Since $X'X$ is $(a+1) \times (a+1)$ and is of rank a, one of its eigenvalues is zero. The columns of V are eigenvectors corresponding to the eigenvalue zero. A condition for estimability is $p'VV' = 0$. Thus, to characterize estimability the equation $p'VV' = 0$ must be solved.

To obtain the eigenvector for the zero eigenvalue observe that

$$
\begin{bmatrix} an & n1'_a \\ n1_a & nI_a \end{bmatrix} \begin{bmatrix} u \\ v_1 \\ \vdots \\ v_a \end{bmatrix} = 0 \tag{10.1.8}
$$

or

$$
anu + n \sum_{i=1}^{a} v_i = 0, \tag{10.1.9}
$$

$$
nu + nv_i = 0, \quad 1 \le i \le a.
$$

Then $u = -1$ and $v_i = 1$. Thus,

$$
V = \begin{bmatrix} -\frac{1}{\sqrt{a+1}} \\ \frac{1a}{\sqrt{a+1}} \end{bmatrix}. \tag{10.1.10}
$$

Then

$$
VV' = \begin{bmatrix} -\frac{1}{\sqrt{a+1}} \\ \frac{1a}{\sqrt{a+1}} \end{bmatrix} \begin{bmatrix} -\frac{1}{\sqrt{a+1}} & \frac{1a'}{\sqrt{a+1}} \end{bmatrix} \tag{10.1.11}
$$

$$
= \begin{bmatrix} \frac{1}{a+1} & -\frac{1a'}{a+1} \\ -\frac{1a}{a+1} & \frac{J}{a+1} \end{bmatrix}.
$$

Let $p' = [c_0, d_1, d_2, \ldots, d_a]$. Then

$$
p'VV' = 0 \quad \text{iff} \quad c_0 - \sum_{i=1}^{a} d_i = 0. \tag{10.1.12}
$$

Condition (10.1.12) then becomes a criterion for estimability. When $c_0 = 0$ (10.1.12) corresponds to the condition defining an orthogonal contrast. The estimable parametric functions are thus of the form

$$p'\hat{\beta} = \left(\sum_{i=1}^{a} d_i\right)\hat{\mu} + \sum_{i=1}^{a} d_i\hat{\alpha}_i$$

and include the orthogonal contrasts. When the coefficient of μ is non-zero an alternate expression for estimable parametric functions is

$$p'\hat{\beta} = \mu + \sum_{i=1}^{a} e_i\hat{\alpha}_i, \quad \text{where} \quad \sum_{i=1}^{a} e_i = 1.$$

Often an experimenter wishes to compare the effect of treatments with that of a control. For example, a medical person might want to compare the effect of three different painkillers with a placebo (e.g., a sugar pill). The contrast of interest would be

$$3a_0 - a_1 - a_2 - a_3 = 0. \qquad (10.1.13)$$

It is easy to see that (10.1.13) is an estimable parametric function.

The Moore-Penrose inverse may be obtained by recalling that for any generalized inverse of $X'X$,

$$U'GU = \Lambda^{-1}. \qquad (10.1.14a)$$

Then

$$UU'GUU' = U\Lambda^{-1}U' = (X'X)^{+}. \qquad (10.1.14b)$$

From (10.1.11)

$$I - VV' = \begin{bmatrix} \frac{a}{a+1} & \frac{1a'}{a+1} \\ \frac{1a}{a+1} & I - \frac{J}{a+1} \end{bmatrix} = UU'. \qquad (10.1.15)$$

Substituting (10.1.6) and (10.1.15) into (10.1.14) it is found that the Moore-Penrose inverse is

$$(X'X)^{+} = \begin{bmatrix} \frac{a}{n(a+1)^2} & \frac{1a'}{(a+1)^2} \\ \frac{1a}{n(a+1)^2} & \frac{1}{n}(I - \frac{(a+2)}{(a+1)^2}J) \end{bmatrix}. \qquad (10.1.16)$$

The estimates of the regression coefficients are

$$
\begin{bmatrix} \hat{\mu} \\ \hat{\alpha} \end{bmatrix} = \begin{bmatrix} \frac{a}{n(a+1)^2} & \frac{1a'}{(a+1)^2} \\ \frac{1a}{n(a+1)^2} & \frac{1}{n}\left(I - \frac{(a+2)}{(a+1)^2}J\right) \end{bmatrix} \begin{bmatrix} Y_{..} \\ Y_{1.} \\ \vdots \\ Y_{a.} \end{bmatrix}. \tag{10.1.17}
$$

Then

$$
\hat{\mu} = \frac{a}{a+1}\bar{Y}_{..}
$$

and

$$
\hat{\alpha}_i = \bar{Y}_i - \frac{a}{a+1}\bar{Y}_{..} . \tag{10.1.18}
$$

Observe that for parametric functions

$$
c_0\mu + \sum_{i=1}^{a} d_i\alpha_i = \sum_{i=1}^{a} d_i Y_{i..} . \tag{10.1.19}
$$

The ANOVA procedure will now briefly be summarized for both the fixed effect and the variance components model. It will be seen that for the variance component model the quantities to be estimated are the expected mean squares. These are estimated by the sums of squares used in the ANOVA table.

In the fixed effect model it is assumed that the α_i are constants and

$$
E(\varepsilon) = 0 \quad \text{and} \quad D(\varepsilon) = \sigma^2 I. \tag{10.1.20}
$$

The usual objective is to test the hypotheses

$$
H_0 : \alpha_1 = \alpha_2 \ldots = \alpha_a \tag{10.1.21}
$$

vs. the alternative that at least one of the above equations is not true.

Let I_a, I_n be $a \times a$ and $n \times n$ dimensional identity matrices. Let J_a and J_n be $a \times a$ and $n \times n$ dimensional matrices of ones. The sum of squares corrected for the mean is

$$\text{TSS} = \sum_{i=1}^{n} \sum_{j=1}^{a} (Y_{ij} - \overline{Y}_{..})^2$$

$$= Y' \left[I_a \otimes I_n - \frac{1}{n}(I_a \otimes J_n) \right] Y$$

$$+ Y' \left[\frac{1}{n} I_a \otimes J_n - \frac{1}{na}(J_a \otimes J_n) \right] Y$$

$$= \sum_{i=1}^{a} \sum_{j=1}^{n} (Y_{ij} - \overline{Y}_{i.})^2 + n \sum_{i=1}^{a} (\overline{Y}_{i.} - \overline{Y}_{..})^2. \quad (10.1.22)$$

The first term in (10.1.19) is the error sum of squares SSE; the second term in (10.1.19) is called the regression, between or treatment sum of squares SSTR. The test of hypothesis in (10.1.17) is performed by forming the ANOVA Table 10.1.

The test statistic is the ratio

$$F = \frac{\frac{\text{SSTR}}{(a-1)}}{\frac{\text{SSE}}{(n-1)a}} \quad (10.1.23)$$

and follows a Snedecor F distribution.

The sums of squares in (10.1.19) will be useful to formulate James-Stein type estimators in the next section.

In the variance component model it is assumed that the α_i are normal random variables with

$$E(\alpha_i) = 0 \quad \text{and} \quad D(\alpha_i) = \sigma_\alpha^2. \quad (10.1.24)$$

Table 10.1

SOURCE	DF	SS	MS
BETWEEN	$a - 1$	SSTR	$\text{SSTR}/(a-1)$
WITHIN	$(n-1)a$	SSE	$\text{SSE}/(n-1)a$
TOTAL	$na - 1$	SST	

Using the assumptions in (10.1.20) and (10.1.24) it is proved in most standard textbooks on experimental design that

$$E\left(\frac{\text{SSE}}{(n-1)a}\right) = \sigma^2 \qquad (10.1.25a)$$

and

$$E\left(\frac{\text{SSTR}}{a-1}\right) = \sigma^2 + n\sigma_\alpha^2. \qquad (10.1.25b)$$

The form of the ANOVA table is the same but the hypothesis being tested is

$$H_0 : \sigma_\alpha^2 = 0 \quad \text{vs.} \quad H_1 : \sigma_\alpha^2 \neq 0.$$

Also the variance components may be estimated by solving the system of equations

$$\hat{\sigma}^2 + n\hat{\sigma}_\alpha^2 = \frac{1}{a-1}\text{SSTR}$$

and

$$\hat{\sigma}^2 = \frac{1}{(n-1)a}\text{SSE}. \qquad (10.1.26)$$

This is called the ANOVA method of estimating variance components. These estimators will be relevant to the formulation of Bayes and empirical Bayes estimators in the next section.

Example 10.1.1. Estimator in an ANOVA model. The data for this example, Table 10.2, is from a large industrial experiment performed at Eastman Kodak Company, Rochester, New York. It was obtained by courtesy of Dr. James Halavin, Associate Professor Department of Mathematics, Rochester Institute of Technology.

Six different units are closer at random from a large number of units of a certain type camera. For each unit the time for the battery pack to recover from the first flash of the camera (i.e., the time from the first flash until the camera's "ready" light went back on) was measured. Six readings were taken for each camera.

The linear model is

$$
Y = \begin{bmatrix}
1_6 & 1_6 & 0 & 0 & 0 & 0 & 0 \\
1_6 & 0 & 1_6 & 0 & 0 & 0 & 0 \\
1_6 & 0 & 0 & 1_6 & 0 & 0 & 0 \\
1_6 & 0 & 0 & 0 & 1_6 & 0 & 0 \\
1_6 & 0 & 0 & 0 & 0 & 1_6 & 0 \\
1_6 & 0 & 0 & 0 & 0 & 0 & 1_6
\end{bmatrix}
\begin{bmatrix}
\mu \\ \alpha_1 \\ \alpha_2 \\ \alpha_3 \\ \alpha_4 \\ \alpha_5 \\ \alpha_6
\end{bmatrix} + \epsilon \,.
$$

The normal equations $X'X\hat{\beta} = X'Y$ are for the above X matrix

$$
\begin{aligned}
36\hat{\mu} + 6\hat{\alpha}_1 + 6\hat{\alpha}_2 + 6\hat{\alpha}_3 + 6\hat{\alpha}_4 + 6\hat{\alpha}_5 + 6\hat{\alpha}_6 &= 188.44, \\
6\hat{\mu} + 6\hat{\alpha}_1 \qquad\qquad\qquad\qquad\qquad\qquad\qquad &= 26.91, \\
6\hat{\mu} \qquad + 6\hat{\alpha}_2 \qquad\qquad\qquad\qquad\qquad\qquad &= 33.21, \\
6\hat{\mu} \qquad\qquad + 6\hat{\alpha}_3 \qquad\qquad\qquad\qquad &= 29.88, \\
6\hat{\mu} \qquad\qquad\qquad + 6\hat{\alpha}_4 \qquad\qquad\qquad &= 29.05, \\
6\hat{\mu} \qquad\qquad\qquad\qquad + 6\hat{\alpha}_5 \qquad &= 36.40, \\
6\hat{\mu} \qquad\qquad\qquad\qquad\qquad + 6\hat{\alpha}_6 &= 32.99.
\end{aligned}
$$

Table 10.2

Camera	1	2	3	4	5	6
1	4.39	5.16	4.89	5.00	6.53	5.71
2	4.34	4.88	4.78	4.45	5.38	4.94
3	4.61	6.81	5.16	5.00	6.21	5.71
4	4.56	6.53	4.94	4.50	5.77	5.71
5	4.89	5.22	5.44	5.54	6.97	6.26
6	4.12	4.61	4.67	4.56	5.54	4.66

Some generalized inverses are

$$
G_1 = \begin{bmatrix}
0 & 0 & 0 & 0 & 0 & 0 & 0 \\
0 & \frac{1}{6} & 0 & 0 & 0 & 0 & 0 \\
0 & 0 & \frac{1}{6} & 0 & 0 & 0 & 0 \\
0 & 0 & 0 & \frac{1}{6} & 0 & 0 & 0 \\
0 & 0 & 0 & 0 & \frac{1}{6} & 0 & 0 \\
0 & 0 & 0 & 0 & 0 & \frac{1}{6} & 0 \\
0 & 0 & 0 & 0 & 0 & 0 & \frac{1}{6}
\end{bmatrix},
$$

$$
G_2 = \begin{bmatrix}
\frac{1}{36} & 0 \\
-\frac{1}{36}I_6 & \frac{1}{6}I_6
\end{bmatrix}
$$

and

$$
G_3 = \begin{bmatrix}
\frac{1}{49} & \frac{1}{294}1_6' \\
\frac{1}{294}1_6 & \frac{1}{6}(I - \frac{8}{49}J)
\end{bmatrix}.
$$

The LS estimators obtained by using G_1 are

$$\hat{\mu} = 0,$$
$$\hat{\alpha}_1 = 4.49,$$
$$\hat{\alpha}_2 = 5.54,$$
$$\hat{\alpha}_3 = 4.98,$$
$$\hat{\alpha}_4 = 4.84,$$
$$\hat{\alpha}_5 = 6.07,$$
$$\hat{\alpha}_6 = 5.50.$$

Using G_2, the LS estimators are

$$\hat{\mu} = 5.23,$$
$$\hat{\alpha}_1 = -.74,$$
$$\hat{\alpha}_2 = .31,$$
$$\hat{\alpha}_3 = -.25,$$
$$\hat{\alpha}_4 = -.39,$$
$$\hat{\alpha}_5 = .84,$$
$$\hat{\alpha}_6 = .27.$$

For the Moore-Penrose inverse G_3,

$$\hat{\mu} = 4.47,$$
$$\hat{\alpha}_1 = 0.00,$$
$$\hat{\alpha}_2 = 1.06,$$
$$\hat{\alpha}_3 = 0.50,$$
$$\hat{\alpha}_4 = 0.36,$$
$$\hat{\alpha}_5 = 1.59,$$
$$\hat{\alpha}_6 = 1.01.$$

Table 10.3

SOURCE	SS	DF	MS	F
Camera Units	9.799	5	1.9598	6.27
Error	9.383	30	0.3128	
Total	19.1825	35		

For all three cases, except for some small differences due to rounding off,

$$\theta_1 = \mu + \hat{\alpha}_1 = 4.5,$$
$$\theta_2 = \mu + \hat{\alpha}_2 = 5.5,$$
$$\theta_3 = \mu + \hat{\alpha}_3 = 5.0,$$
$$\theta_4 = \mu + \hat{\alpha}_4 = 4.8,$$
$$\theta_5 = \mu + \hat{\alpha}_5 = 6.1,$$
$$\theta_6 = \mu + \hat{\alpha}_6 = 5.5.$$

The linear contrasts are all expressible as linear combinations of $\theta_1, \theta_2, \theta_3, \theta_4, \theta_5$, and θ_6. They should be the same for all three cases. The ANOVA table is given in Table 10.3.

The estimates of the variance components are

$$\hat{\sigma}_\alpha^2 = \frac{1.9598 - .3128}{6} = .2745,$$

$$\hat{\sigma}^2 = 0.3128.$$

Observe that the null hypothesis

$$H_0 : \sigma_\alpha^2 = 0$$

is rejected at $\alpha = .05$. Thus, there is significant variability amongst camera units.

Exercise 10.1.1. Show that G_1, G_2 and G_3 in Example 10.1.1 are generalized inverses of $X'X$. Which of the Penrose conditions are satisfied by each one?

Exercise 10.1.2. For the random effects model

$$Y_{ij} = \mu + \alpha_i + \varepsilon_{ij}, \quad 1 \leq i \leq a, \quad 1 \leq j \leq b,$$

where

$$\alpha_i \sim N(0, \sigma_\alpha^2) \quad \text{and} \quad \varepsilon_{ij} \sim N(0, \sigma^2),$$

the α_is and ε_{ij} are all made independent, what is the variance co-variance matrix of Y_{ij}, $1 \leq i \leq a$, $1 \leq j \leq n$?

Exercise 10.1.3. Five sophomore mathematics students at six randomly selected universities were given a standardized Calculus test. Their scores were

University A	85	82	86	85	78
B	81	84	76	79	81
C	90	65	82	84	76
D	90	94	88	85	80
E	65	73	78	80	72
F	55	48	67	75	69

A. Find two sets of least square estimates.
B. Perform the ANOVA and estimate the variance components.
C. Estimate $\hat{\alpha}_1 - \hat{\alpha}_3$.

The above preliminary material will now be used as a basis for formulating the Bayes and the empirical Bayes estimator.

10.2 The Bayes and Empirical Bayes Estimators

The assumptions in (10.1.21) will now be regarded as prior information. The objective will be to obtain Bayes and empirical Bayes estimators for $p'\alpha$ and to study their MSE properties. Thus, assume

$$E(\alpha) = 0 \quad \text{and} \quad D(\alpha) = \sigma_a^2 I \qquad (10.2.1a)$$

and

$$E(\varepsilon|\alpha) = 0 \quad \text{and} \quad D(\varepsilon|\alpha) = \sigma^2 I. \qquad (10.2.1b)$$

Rewrite (10.1.1) in the form

$$Y - (1_a \otimes 1_n)\mu = (I_a \otimes 1_n)\alpha + \varepsilon. \qquad (10.2.2)$$

For each μ the BE

$$p'\hat{\beta} = p'(Z'Z + \sigma^2 I)^{-1}Z'(Y - (1_a \otimes 1_n)\mu). \qquad (10.2.3a)$$

Estimate μ by $\overline{Y}_{..}$. Then

$$p'\hat{\alpha} = p'(Z'Z + \sigma^2 I)^{-1}Z'(Y - (1_a \otimes 1_n))\overline{Y}_{..}). \qquad (10.2.3b)$$

Componentwise,

$$\hat{\alpha}_{1b} = \frac{\sigma_\alpha^2}{\sigma^2 + n\sigma_\alpha^2}(\overline{Y}_{i.} - \overline{Y}_{..})$$

$$= \left(1 - \frac{\sigma^2}{\sigma^2 + n\sigma_\alpha^2}\right)(\overline{Y}_{i.} - \overline{Y}_{..})$$

$$= \left(1 - \frac{\sigma^2}{\sigma^2 + n\sigma_\alpha^2}\right)\hat{\alpha}_i. \qquad (10.2.4)$$

The EBE will be found by subsituting unbiased estimators for the numerator and denominator in the fraction

$$f = \frac{\sigma^2}{\sigma^2 + n\sigma_\alpha^2}.$$

These estimators will be the terms corresponding to the MS in the error sum of squares and treatment sum of squares respectively. Thus

$$\hat{\sigma}^2 = \frac{1}{(n-1)a}\sum_{i=1}^{a}\sum_{j=1}^{n}(Y_{ij} - \overline{Y}_{i.})^2 \qquad (10.2.5)$$

and

$$(\widehat{\sigma^2 + n\sigma_\alpha^2}) = \frac{\hat{\alpha}'(Z'Z)\hat{\alpha}}{a-1}$$

$$= \frac{n\sum_{i=1}^{a}(\bar{Y}_i - \bar{Y}_{..})^2}{(a-1)}. \qquad (10.2.6)$$

Substituting (10.2.5) and (10.2.6) in (10.2.4) yields the EBE

$$\hat{\alpha}_{ebi} = \left(1 - \frac{c(a-1)\hat{\sigma}^2}{a'(Z'Z)a}\right)\hat{\alpha}_i. \tag{10.2.7}$$

The constant c is chosen so that MSE is minimized.

The average MSE of $p'\alpha$ for the estimable parametric functions is

$$p'E(\hat{\alpha}_{eb} - \alpha)(\hat{\alpha}_{eb} - \alpha)'p$$

$$= \frac{p'Ip}{n}\sigma^2 - \frac{2c}{n}p'Tp\frac{\sigma^4}{\sigma^2 + n\sigma_a^2}$$

$$+ \frac{[a(n-1)+2](a-1)^2c^2}{a(a-3)n(n-1)}p'Tp\frac{\sigma^4}{\sigma^2 + n\sigma_a^2}, \tag{10.2.8}$$

where $T = I - (1/a)J$.

Optimum

$$c = \frac{(a-3)(n-1)a}{(a-1)[(n-1)a+2]}. \tag{10.2.9}$$

For the optimum c the MSE of $p'\alpha$ is, by substitution of (10.2.9) into (10.2.8)

$$p'E(\hat{\alpha}_{eb} - \alpha)(\hat{\alpha}_{eb} - \alpha)'p \tag{10.2.10}$$

$$= \frac{p'Ip\sigma^2}{n} - \frac{(a-3)(n-1)}{(n-a)[(n-1)a+2]}\frac{p'Tp}{n}\frac{\sigma^4}{\sigma^2 + n\sigma_a^2}.$$

See the appendix for an outline of the neccessary computation.

Example 10.2.1. James-Stein Estimator. The JS estimator of $\mu + \alpha_i$ is

$$\widehat{\mu + \alpha_i} = \bar{Y}_{..} + \left(1 - \frac{(a-3)(n-1)a}{(a-1)[(n-1)a+2]}\frac{\text{MSE}}{\text{MSA}}\right)(\bar{Y}_{i.} - \bar{Y}_{..}).$$

Then the JS estimator is, when $n = 6$ and $a = 6$,

$$\widehat{\mu + \alpha_i} = \bar{Y}_{..} + \left(1 - \frac{9}{16} \frac{\text{MSE}}{\text{MSA}}\right)(\bar{Y}_{i.} - \bar{Y}_{..}) .$$

For Example 10.1.1

$$\widehat{\mu + \alpha_i} = \bar{Y}_{..} + (.91)(\bar{Y}_{i.} - \bar{Y}_{..}) .$$

Thus the JS estimates are

$$\widehat{\mu + \alpha_1} = 4.55,$$
$$\widehat{\mu + \alpha_2} = 5.51,$$
$$\widehat{\mu + \alpha_3} = 5.00,$$
$$\widehat{\mu + \alpha_4} = 4.88,$$
$$\widehat{\mu + \alpha_5} = 5.99,$$
$$\widehat{\mu + \alpha_6} = 5.47.$$

They exhibit a shrinkage of the treatment means towards the grand mean of 5.23.

Observe that if (MSE/MSA) were larger there would be more shrinkage, e.g., if the hypothesis $H_0 : \sigma_\alpha^2 = 0$ could not be rejected. A pretest estimator is one where for some number c

$$\widehat{\mu + \alpha_i} = \begin{cases} \bar{Y}_{..} + \hat{\alpha}_i^{eb} & \text{if } F = \frac{\text{MSA}}{\text{MSE}} < c, \\[2mm] \bar{Y}_{i.} & \text{if } F = \frac{\text{MSA}}{\text{MSE}} > c. \end{cases}$$

Exercise 10.2.1. For the data of Exercise 10.1.1 find the JS estimate.

Exercise 10.2.2. For the one way ANOVA model explain why the JS estimator cannot be formulated when $a < 3$.

Exercise 10.2.3. For the one way ANOVA find

$$m = E\left[\sum_{i=1}^{a}(\alpha_i^{eb} - \alpha_i)^2\right]$$

without averaging over the prior for optimum c.

Exercise 10.2.4. For the EBE find the relative savings loss

$$\text{RSL} = \frac{\text{MSE(EBE)} - \text{MSE(BE)}}{\text{MSE(LS)} - \text{MS(BE)}}.$$

Get numerical values for the data of Example 10.2.1 and Exercise 10.2.1. The RSL is a measure of the efficiency of the EBE as compared with the BE. It was proposed by Efron and Morris (1973).

10.3 The Two Way Classification

The linear model under consideration is

$$Y = [1_a \otimes 1_b \ I_a \otimes 1_b \ 1_a \otimes I_b]\gamma + \varepsilon, \qquad (10.3.1)$$

where

$$\gamma = [\mu, \alpha, \beta] = [\mu, \alpha_1, \alpha_2, \dots, \alpha_a, \beta_1, \beta_2, \dots, \beta_b].$$

The normal equations necessary to obtain the LS estimators are

$$\begin{bmatrix} ab & b1_{a'} & a1'_b \\ b1_a & bI_a & J_{a\times b} \\ a1_b & J_{b\times a} & aI_b \end{bmatrix} \begin{bmatrix} \mu \\ \alpha \\ \beta \end{bmatrix} = \begin{bmatrix} Y_{..} \\ Y_{1.} \\ Y_{2.} \\ \vdots \\ Y_{a.} \\ Y_{.1} \\ Y_{.2} \\ \vdots \\ Y_{.b} \end{bmatrix}. \qquad (10.3.2)$$

Given the constraints

$$\sum_{i=1}^{a}\alpha_i = 0, \qquad \sum_{j=1}^{b}\beta_j = 0, \qquad (10.3.3)$$

the LS estimators are

$$\begin{aligned}
\hat{\mu} &= \bar{Y}_{..}, \\
\hat{\alpha}_i &= \bar{Y}_{i.} - \bar{Y}_{..}, \\
\hat{\beta}_j &= \bar{Y}_{.j} - \bar{Y}_{..}.
\end{aligned} \qquad (10.3.4)$$

Alternatively, the least square estimator may be written

$$\begin{bmatrix} \hat{\mu} \\ \hat{\alpha}_1 \\ \vdots \\ \hat{\alpha}_a \\ \hat{\beta}_1 \\ \vdots \\ \hat{\beta}_b \end{bmatrix} = GX'Y, \qquad (10.3.5)$$

where

$$G = \begin{bmatrix} \frac{1}{ab} & 0 & 0 \\ -\frac{1}{ab}1_a & \frac{1}{b}I_a & 0 \\ -\frac{1}{ab}I_b & 0 & \frac{1}{a}I_b \end{bmatrix}. \qquad (10.3.6)$$

Estimability is characterized by finding the form of vectors

$$p' = [l_0, m_1, m_2, \ldots, m_a, n_1, \ldots, n_b] \qquad (10.3.7a)$$

that satisfy

$$p'VV' = 0. \tag{10.3.7b}$$

The columns of the matrix V consist of the eigenvectors of 0 for the $X'X$ matrix, e.g.,

$$c'X'X = 0, \tag{10.3.8a}$$

where

$$c = [s_0, w_1, w_2, \ldots, w_a, t_a, t_2, \ldots, t_b] \tag{10.3.8b}$$

or equivalently,

$$abs_0 + b\sum_{i=1}^{a} w_i + a\sum_{j=1}^{b} w_j = 0,$$

$$as_0 + bw_i + \sum_{j=1}^{b} w_j = 0, \quad 1 \le i \le a, \tag{10.3.8c}$$

$$as_0 + \sum_{j=1}^{b} w_j + aw_j = 0, \quad 1 \le j \le b.$$

The matrix

$$V = \begin{bmatrix} 0 & \frac{2}{\sqrt{2+a+b}} \\ \frac{1a}{\sqrt{a+b}} & \frac{-1a}{\sqrt{2+a+b}} \\ \frac{-1b}{\sqrt{a+b}} & \frac{-1b}{\sqrt{2+a+b}} \end{bmatrix} \tag{10.3.9}$$

and

$$VV' = \begin{bmatrix} \dfrac{4}{2+a+b} & \dfrac{-2 \cdot 1 a}{2+a+b} & \dfrac{-2 \cdot 1 b'}{2+a+b} \\[2ex] \dfrac{-2 \cdot 1 a}{2+a+b} & \dfrac{2(1+a+b)}{(a+b)(2+a+b)} J_{a \times a} & \dfrac{-2}{(2+a+b)(a+b)} J_{a \times b} \\[2ex] \dfrac{-2 \cdot 1 b}{2+a+b} & \dfrac{-2 J_{b \times a}}{(a+b)(2+a+b)} & \dfrac{2(a+a+b)}{(a+b)(2+a+b)} J_{b \times b} \end{bmatrix} . \quad (10.3.10)$$

Let $p' = [l_0, m_1, \ldots, m_a, n_1, \ldots, n_b]$. The estimable parametric functions satisfy (10.3.7). Equation (10.3.7) may now be written

$$2l_0 - \sum_{i=1}^{a} m_i - \sum_{j=1}^{b} n_j = 0,$$

$$(a+b)l_0 - (1+a+b)\sum_{i=1}^{a} m_j + \sum_{j=1}^{b} n_j = 0, \qquad (10.3.11)$$

$$(a+b)l_0 + \sum_{i=1}^{a} m_i - (1+a+b)\sum_{j=1}^{b} n_j = 0 .$$

The parametric functions then are contrasts if $l_0 = 0$. If $l_0 \neq 0$ the solution to (10.3.11) is

$$\sum_{i=1}^{a} m_i = l_0 = \sum_{i=1}^{b} n_i . \qquad (10.3.12)$$

The estimable functions are linear combinations of $u + c'a + d'\beta$, where $c'c = 1$ and $d'd = 1$.

The estimable parametric functions include the linear contrasts $c'\alpha$ and $d'\beta$ but not the individual parameters.

The ANOVA for the randomized block design will now be summarized.

Using notation similar to that in equation (10.1.22) the sum of squares corrected for the mean is

$$\text{TSS} = \sum_{i=1}^{a}\sum_{j=1}^{b}(Y_{ij} - \overline{Y}_{..})^2$$

$$= Y'\left[I_a \otimes I_b - \frac{1}{ab}(J_a \otimes J_b)\right]Y$$

$$= Y'\left[I_a \otimes I_b - \frac{1}{b}I_a \otimes J_b - \frac{1}{a}J_a \otimes I_b + \frac{1}{ab}J_a \otimes J_b\right]Y$$

$$+ Y'\left[\frac{1}{b}I_a \otimes I_b - \frac{1}{ab}(J_a \otimes J_b)\right]Y$$

$$+ Y'\left[\frac{1}{a}(J_a \otimes I_b) - \frac{1}{ab}(J_a \otimes J_b)\right]Y$$

$$= \sum_{i=1}^{a}\sum_{j=1}^{b}(Y_{ij} - \overline{Y}_{i.} - \overline{Y}_{.j} + \overline{Y}_{..})^2$$

$$+ b\sum_{i=1}^{a}(\overline{Y}_{i.} - \overline{Y}_{..})^2 + a\sum_{j=1}^{b}(\overline{Y}_{.j} - \overline{Y}_{..})^2. \qquad (10.3.13)$$

The first term in (10.3.13) is the error sum of squares. The second term is a treatment sum of squares. The third term is a treatment sum of squares if the experimental design is a factorial design without interaction. If the experimental design is a randomized block design the third term is a block sum of squares.

The usual ANOVA table for the randomized block design is displayed in Table 10.4.

Table 10.4

SOURCE	DF	SS	MS
Treatment	$a-1$	SSTR	SSTR$/(a-1)$ =MSTR
Block	$b-1$	SSBL	SSBL$/(b-1)$=MSBL
Error	$(a-1)(b-1)$	SSE	SSE$/(a-1)(b-1)$=MSE
Total	$ab-1$		

In the fixed effect model the hypothesis being tested are

$$H_0 : \alpha_0 = \alpha_2 \cdots = \alpha_a. \tag{10.3.14a}$$

Using the test statistics

$$F = \frac{\frac{SSTR}{(a-1)}}{\frac{SSE}{(a-1)(b-1)}} \sim F_{(a-1),(a-1)(b-1)} \tag{10.3.14b}$$

and

$$H_0 : \beta_1 = \beta_2 \ldots = \beta_a, \tag{10.3.15a}$$

$$F = \frac{\frac{SSBL}{(b-1)}}{\frac{SSE}{(a-1)(b-1)}} \sim F_{(b-1),(a-1)(b-1)}. \tag{10.3.15b}$$

The other two situations will be of importance here, the random effects model where α and β parameters are random variables and the mixed model where one of the parameters for one factor is a random variable and the parameter for the other factor is constant.

For the random effects model assume

$$\alpha_i \sim N(0,\sigma_\alpha^2), \quad \beta_i \sim N(0,\sigma_b^2). \tag{10.3.16a}$$

Now

$$\begin{aligned} E(MSA) &= \sigma^2 + b\sigma_a^2, \\ E(MSB) &= \sigma^2 + a\sigma_b^2, \\ E(MSE) &= \sigma^2 \ . \end{aligned} \tag{10.3.16b}$$

The hypotheses being tested using the same F statistics as before is

$$H_0 : \sigma_\alpha^2 = 0 \tag{10.3.17}$$

and

$$H_0 : \sigma_\beta^2 = 0.$$

When, say, the α_i are fixed and the β_i are random variables the hypothesis being tested is

$$H_0 : \alpha_i = 0, \quad 1 \le i \le s, \qquad (10.3.18)$$

and

$$H_0 : \sigma_\alpha^2 = 0.$$

The

$$E(\text{MSA}) = \frac{b \sum_{i=1}^{a} \alpha_i^2}{a - 1} + \sigma^2,$$

$$E(\text{MSB}) = +a\sigma_\alpha^2 + \sigma^2, \qquad (10.3.19)$$

$$E(\text{MSE}) = \sigma^2 .$$

Example 10.3.1. Estimating Variance Components Suppose that in Example 10.1.1 the readings were obtained for six different battery types chosen randomly from a large group of battery types. In Table 10.1 the rows are battery types and the columns are the limits. The ANOVA table is given in Table 10.5.

Table 10.5

SOURCE	DF	SS	MS	EMS	F
Units	5	9.7994	1.9599	$\sigma^2 + 6\sigma_\alpha^2$	11.53
Battery	5	5.1335	1.0267	$\sigma^2 + 6\sigma_\beta^2$	6.04
Error	25	4.2495	.1700		
Total	35	19.1825			

Both the hypotheses

$$H_0 : \sigma_\alpha^2 = 0$$

and

$$H_0 : \sigma_\beta^2 = 0$$

are rejected at $\alpha = .05$ and $\alpha = .01$.

Exercise 10.3.1. Estimate the variance components in Example 10.3.1.

Exercise 10.3.2. Write the linear model for the two way ANOVA when $a = 6$ and $b = 6$, and find the LS estimates for

$$\mu + \alpha_i, \quad i = 1, 2, 3, 4, 5, 6,$$

$$\mu + \beta_j, \quad j = 1, 2, 3, 4, 5, 6.$$

Does your answer depend on your choice of solution to the normal equations? Find a basis for the column space of X.

Exercise 10.3.3. Consider the model

$$Y_{ij} = \mu + \alpha_i + \beta_j + \epsilon_{ij}, \quad 1 \le i \le a, \, 1 \le j \le b,$$

where $\alpha_i, \beta_j, \epsilon_{ij}$ are independent and satisfy assumptions (10.3.16a). What is the variance-covariance matrix of Y_{ij}?

10.4 The Bayes and Empirical Bayes Estimators

The formulation of the BE and EBE for the two way ANOVA model is a bit harder than for the one way case. For the linear model

$$Y = (1_a \otimes 1_b)\mu + (I_a \otimes 1_b)\alpha + (1_a \otimes I_b)\beta + \varepsilon \qquad (10.4.1)$$

the assumptions are

$$E(\varepsilon|\alpha,\beta) = 0, \quad D(\varepsilon|\alpha,\beta) = \sigma^2 I. \tag{10.4.2}$$

The prior assumptions on α and β are

$$E(\alpha) = 0, \quad D(\alpha) = \sigma_\alpha^2 I, \tag{10.4.3}$$

$$E(\beta) = 0, \quad D(\beta) = \sigma_\beta^2 I. \tag{10.4.4}$$

The parameters α and β are assumed to be independent.
To formulate the BE use the model (10.4.1) in the form

$$Y - (1_a \otimes 1_b)\mu = Z\begin{bmatrix} \alpha \\ \beta \end{bmatrix} + \varepsilon, \tag{10.4.5}$$

where

$$Z = [I_a \otimes 1_b \quad 1_a \otimes I_b]. \tag{10.4.6}$$

As a result the form of the BE is

$$p'\beta = p'(Z'Z + \sigma^2 F^{-1})^{-1} Z'(Y - (1_a \otimes 1_b)\mu), \tag{10.4.7}$$

where

$$F = \begin{bmatrix} \sigma_\alpha^2 I & 0 \\ 0 & \sigma_\beta^2 I \end{bmatrix}. \tag{10.4.8}$$

Using results on patterned matrices (see (2.6.2))

$$(Z'Z + \sigma^2 F^{-1})^{-1} = \begin{bmatrix} \dfrac{(b\sigma_\alpha^2 + \sigma^2)}{\sigma_\alpha^2} I & J_{b\times a} \\ J_{a\times b} & \dfrac{(a\sigma_\beta^2 + \sigma^2)}{\sigma_\beta^2} I \end{bmatrix}^{-1} \tag{10.4.9}$$

$$= \begin{bmatrix} \dfrac{1}{a_1} I + b_1 J & b_2 J \\ b_2 J & \dfrac{1}{a_3} I + b_2 J \end{bmatrix},$$

where

$$a_1 = \frac{(b\sigma_\alpha^2 + \sigma^2)}{\sigma_\alpha^2}, \quad a_2 = 1 \quad \text{and} \quad a_3 = \frac{(a\sigma_\beta^2 + \sigma^2)}{\sigma_\beta^2}. \qquad (10.4.10)$$

Also

$$b_1 = \frac{-b}{a_1(ab - a_3 a_1)}, \quad b_2 = \frac{1}{ab - a_3 a_1} \quad \text{and} \quad b_3 = \frac{-a}{a_1(ab - a_3 a_1)}. \qquad (10.4.11)$$

The BE in (10.4.7) may be written

$$\hat{\alpha}_i = \frac{b}{a_1}(\bar{Y}_{i.} - \mu) + \frac{(b_1 + b_2)}{ab}(\bar{Y}_{..} - \mu) \qquad (10.4.12a)$$

and

$$\hat{\beta}_j = \frac{a}{a_3}(\bar{Y}_{.j} - \mu) + \frac{(b_2 + b_3)}{ab}(\bar{Y}_{..} - \mu) . \qquad (10.4.12b)$$

When μ is unknown (usually the case in a practical situation), (10.4.12) becomes

$$\hat{\alpha}_i = \frac{b}{a_1}(\bar{Y}_{i.} - \bar{Y}_{..}) = \frac{b\sigma_a^2}{b\sigma_\alpha^2 + \sigma^2}(\bar{Y}_{i.} - \bar{Y}_{..}) = \left(1 - \frac{\sigma^2}{b\sigma_\alpha^2 + \sigma^2}\right)(\bar{Y}_{i.} - \bar{Y}_{..}) \qquad (10.4.13a)$$

and

$$\hat{\beta}_j = \frac{a}{a_3}(\bar{Y}_{.j} - \bar{Y}_{..}) = \frac{a\sigma_b^2}{a\sigma_b^2 + \sigma^2}(\bar{Y}_{.j} - \bar{Y}_{..}) = \left(1 - \frac{\sigma^2}{a\sigma_b^2 + \sigma^2}\right)(\bar{Y}_{.j} - \bar{Y}_{..}). \qquad (10.4.13b)$$

When σ^2, σ_a^2 are unknown σ^2 may be estimated using the within sum of squares divided by $(a-1)(b-1)$:

$$\hat{\sigma}^2 = \frac{1}{(a-1)(b-1)} \sum_{i=1}^{a} \sum_{j=1}^{b} (Y_{ij} - \bar{Y}_{i.} - \bar{Y}_{.j} + \bar{Y}_{..})^2. \qquad (10.4.14a)$$

Also

$$b\hat{\sigma}_a^2 + \hat{\sigma}^2 = \frac{b\sum_{i=1}^{a}(\bar{Y}_{i.} - \bar{Y}_{..})^2}{(a-1)} \qquad (10.4.14b)$$

and

$$a\hat{\sigma}_b^2 + \hat{\sigma}^2 = \frac{a\sum_{j=1}^{b}(\bar{Y}_{.j} - \bar{Y}_{..})^2}{(b-1)}. \qquad (10.4.14c)$$

Replace the functions of the parameter values in (10.4.13) by the estimators in (10.4.14). Then for the c that produces the smallest MSE,

$$\hat{\alpha}_{ebi} = \left(1 - \frac{c\hat{\sigma}^2(b-1)}{a\sum_{i=1}^{a}(\bar{Y}_{i.} - \bar{Y}_{..})^2}\right)(\bar{Y}_{i.} - \bar{Y}_{..}) \qquad (10.4.15a)$$

and

$$\hat{\beta}_{ebj} = \left(1 - \frac{c\hat{\sigma}^2(a-1)}{b\sum_{j=1}^{a}(\bar{Y}_{.j} - \bar{Y}_{..})^2}\right)(\bar{Y}_{.j} - \bar{Y}_{..}). \qquad (10.4.15b)$$

This is the James-Stein type empirical Bayes estimator.

These estimators are the same as the ones that would have been obtained from the models

$$Y = (1_a \otimes 1_b)\mu + (I_a \otimes 1_b)\alpha + \varepsilon \qquad (10.4.16)$$

and

$$Y = (1_a \otimes 1_b)\mu + (I_a \otimes 1_b)\beta + \varepsilon \qquad (10.4.17)$$

where

$$\alpha \sim N(0, \sigma_\alpha^2 I), \beta \sim N(0, \sigma_\beta^2 I) \qquad (10.4.18)$$

and

$$\varepsilon \sim N(0, \sigma^2 I). \qquad (10.4.19)$$

Thus the MSE in (10.4.20) can be obtained by the method of computation used for the one way ANOVA model.

The MSE of $p'\hat{\alpha}$ is for estimable parametric functions

$$p'E(\hat{\alpha}_{eb} - \alpha)'(\hat{\alpha}_{eb} - \alpha)p$$

$$=\frac{p'Ip}{b}\sigma^2 - \frac{2c}{b}\frac{\sigma^2}{\sigma^2 + b\sigma_\alpha^2}T_1 \qquad (10.4.20)$$

$$+ \frac{[(a-1)(b-1)+2](a-1)c^2}{a(a-3)b(b-1)}\frac{\sigma^2}{\sigma^2 + b\sigma_\alpha^2}T_1$$

with $T_1 = I - (1/a)J$.

Optimum

$$c = \frac{a(a-3)(b-1)}{[(a-1)(b-1)+2](a-1)} . \qquad (10.4.21)$$

For optimum c the

$$\text{MSE} = \frac{p'Ip}{b}\sigma^2 - \frac{a(a-3)(b-1)}{[(a-1)(b-1)+2](a-1)}\frac{p'T_1p}{b} . \qquad (10.4.22)$$

For $p'\hat{\beta}$,

$$p'E(\hat{\beta}eb - \beta)'(\hat{\beta}eb - \beta)p$$

$$=\frac{p'Ip}{a}\sigma^2 - \frac{2c}{a}\frac{\sigma^2}{\sigma^2 + a\sigma_\beta^2}T_2 \qquad (10.4.23)$$

$$+ \frac{[(a-1)(b-1)+2](b-1)c^2}{b(b-3)a(a-1)}\frac{\sigma^2}{\sigma^2 + a\sigma_\beta^2}T_2$$

with $T_2 = I - (1/b)J$.

Optimum

$$c = \frac{b(b-3)(a-1)}{[(b-1)(a-1)+2](b-1)} \cdot \qquad (10.4.24)$$

For optimum c,

$$\text{MSE} = \frac{p'Ip}{a}\sigma^2 - \frac{b(b-3)(a-1)}{[(b-1)(a-1)+2](b-1)}\frac{p'T_2 p}{\sigma^2 + a\sigma_\beta^2} \cdot \qquad (10.4.25)$$

Exercise 10.4.1. For the design in Example 10.3.1 find the JS estimates of

$$\mu + \alpha_i, \quad i = 1,2,3,4,5,6,$$

$$\mu + \beta_j, \quad j = 1,2,3,4,5,6.$$

Exercise 10.4.2. Find the conditional MSE for the two way ANOVA, i.e.,

$$M_1 = E_\alpha \left[\sum_{i=1}^{a} (\alpha_i^{JS} - \alpha_i)^2 \right],$$

$$M_2 = E_\beta \left[\sum_{j=1}^{b} (\beta_j^{JS} - \beta_j)^2 \right].$$

Exercise 10.4.3. Consider the model

$$Y_{ijk} = \mu + \alpha_i + \beta_j + \gamma_k + \epsilon_{ijk}$$

when

$$i = 1,2\ldots p,$$
$$j = 1,2\ldots p,$$
$$k = 1,2\ldots p$$

and

$$\alpha_i \sim N(0,\sigma_\alpha^2), \quad \beta_j \sim N(0,\sigma_\beta^2), \quad \gamma_k \sim N(0,\sigma_\gamma^2).$$

A. Find the LS estimators.
B. Formulate the BE and JS estimators.
C. What is the form of the estimable parametric functions?
D. Find the average MSE.
E. Can the JS estimators be formulated when $p < 3$?

Exercise 10.4.4. Find the singular value decomposition of the X matrix in (10.3.1). (This is a messy problem.)

Exercise 10.4.5. Formulate JS estimators for the random component in a two way ANOVA mixed model, i.e., one fixed factor-one random factor.

Exercise 10.4.6. Consider the linear models
A.
$$Y_{ijk} = \mu + \tau_i + \beta_{ji} + \epsilon_{ijk},$$
$$\tau_i \sim N(0,\sigma_\tau^2), \quad \beta_{j(i)} \sim N(0,\sigma_\beta^2), \quad \epsilon_{ijk} \sim N(0,\sigma^2),$$
$$1 \le i \le a, \quad 1 \le j \le b, \quad 1 \le k \le n;$$

B.
$$Y_{ijk} = \mu + \tau_i + \beta_j + (\tau\beta)_{ij} + \epsilon_{ijk},$$
$$\tau_i \sim N(0,\sigma_\tau^2), \quad \beta_j \sim N(0,\sigma_\beta^2), \quad (\tau\beta)_{ij} \sim N(0,\sigma_{\tau\beta}^2),$$
$$1 \le i \le a, \quad 1 \le j \le b, \quad 1 \le k \le n.$$

(1) For the models A and B respectively show that

$$\theta_{ij} = \mu + \tau_i + \beta_{j(i)}$$

and

$$\theta_{ij} = \mu + \tau_i + \beta_j + (\tau\beta)_{ij}$$

contain bases for the vector space of estimable parametric functions.

(2) Formulate the BE and JS estimators for $\hat{\theta}_{ij}$ by first finding the BE for τ, β, etc.

(3) Obtain their optimal MSE averaging over the prior distribution.

10.5 Summary

Bayes and empirical Bayes estimators are formulated for the estimable parametric function in a one and two way ANOVA model. Estimates of linear contrasts are obtained with average MSE less than the usual least square estimates used for the analysis of variance.

Appendix to Section 10.2

Calculation of MSE *of Section* 10.2

Assume the prior distribution and the data are normally distributed.

Observe that the MSE of $p'\hat{\alpha}_{eb}$ is the sum of the three terms as shown in (A2.1) below. The expectations in these individual terms must then be obtained. Thus, for the estimable parametric functions,

$$p'E(\hat{\alpha}_{eb} - \alpha)(\hat{\alpha}_{eb} - \alpha)'p$$

$$=p'E(\hat{\alpha} - \alpha)(\hat{\alpha} - \alpha)'p - 2c(a-1)p'Ep\left[\frac{\hat{\sigma}^2\hat{\alpha}(\hat{\alpha} - \alpha)'}{\alpha'(Z'Z)\alpha}\right]p$$

$$+ c^2(a-1)^2 p'E\left[\frac{\hat{\sigma}^4\hat{\alpha}\hat{\alpha}'}{[\alpha(Z'Z)\alpha]^2}\right]p'. \tag{A2.1}$$

Since $p'\hat{\alpha}$ is MVUE for $p'\alpha$ and is a linear combination of $\overline{Y}_{j.}$ its MSE, the first term of (A2.1) is simply

$$l_1 = \frac{p'Gp\sigma^2}{n} = \frac{p'Ip}{n}\sigma^2.$$

To calculate the second term first condition the expectation on the sufficient statistics $\hat{\sigma}^2$ and $\hat{\alpha}$. Thus,

$$E\left[\frac{\hat{\sigma}^2\hat{\alpha}(\hat{\alpha} - \alpha)'}{\hat{\alpha}'(Z'Z)\hat{\alpha}}\right] = EE_{\hat{\sigma}^2,\hat{\alpha}}\left[\frac{\hat{\alpha}(\hat{\alpha} - \alpha)'\hat{\sigma}^2}{\hat{\alpha}'(Z'Z)\hat{\alpha}}\right]$$

$$= E\left[\frac{\hat{\sigma}^2\hat{\alpha}\hat{\alpha}'}{\hat{\alpha}(Z'Z)\hat{\alpha}}\right] - E\left[\frac{\hat{\sigma}^2\hat{\alpha}}{\hat{\alpha}'(Z'Z)\hat{\alpha}}E_{\hat{\alpha}}(\alpha)'\right]. \tag{A2.2}$$

The sufficiency principle (see Lindley (1965)), states that the distribution of a parameter θ given a sufficient statistic $t(x)$ where $x = (x_1, \cdots, x_n)$ is the same as the posterior distribution of θ, i.e.,

$$p(\theta|t(x)) = p(\theta|x). \tag{A2.3}$$

Thus $E_{\hat{\alpha}}(\alpha)$ is the mean of the posterior distribution, the Bayes estimator. Consequently,

$$
\hat{\alpha}_{ib} = E_\alpha(\alpha_i) = \left(1 - \frac{\sigma^2}{\sigma^2 + n\sigma_\alpha^2}\right)(\bar{Y}_{i.} - \mu)
$$

$$
= \left(1 - \frac{\sigma^2}{\sigma^2 + n\sigma_\alpha^2}\right)(\bar{Y}_{i.} - \bar{Y}_{..} + \bar{Y}_{..} - \mu)
$$

$$
= \left(1 - \frac{\sigma^2}{\sigma^2 + n\sigma_\alpha^2}\right)(\hat{\alpha}_i + \bar{Y}_{..} - \mu) . \qquad (A2.4)
$$

Thus,

$$
\alpha_i - \hat{\alpha}_{ib} = \alpha_i - \left(1 - \frac{\sigma^2}{\sigma^2 + n\sigma_\alpha^2}\right)(\hat{\alpha}_i + \bar{Y}_{..} - \mu)
$$

$$
= -(\bar{Y}_{..} - \mu) + \frac{\sigma^2}{\sigma^2 + n\sigma_\alpha^2}(\hat{\alpha}_i + \bar{Y}_{..} - \mu). \qquad (A2.5)
$$

From (A2.2) and (A2.5),

$$
E\left[\frac{\hat{\sigma}^2 \hat{\alpha}(\alpha - \hat{\alpha})'}{\hat{\alpha}'(Z'Z)\hat{\alpha}}\right]
$$

$$
= -\sigma^2 E\left[\frac{\hat{\alpha}(\bar{Y}_{..} - \mu)}{\hat{\alpha}'(Z'Z)\hat{\alpha}}\right] + \frac{\sigma^4}{\sigma^2 + n\sigma_\alpha^2} E\left[\frac{\hat{\alpha}(\hat{\alpha} + \bar{Y} - \mu)}{\alpha'(Z'Z)\hat{\alpha}}\right] \qquad (A2.6)
$$

$$
= \frac{\sigma^4}{\sigma^2 + n\sigma_\alpha^2} E\left[\frac{\hat{\alpha}\hat{\alpha}'}{\hat{\alpha}'(Z'Z)\hat{\alpha}}\right] - \frac{n\sigma_\alpha^2\sigma^2}{\sigma^2 + n\sigma_\alpha^2} E\left[\frac{\hat{\alpha}(\bar{Y}_{..} - \mu)}{\hat{\alpha}'(Z'Z)\hat{\alpha}}\right].
$$

It is easily seen that

$$
\sum_{i=1}^{a}(\bar{Y}_{i.} - \bar{Y}_{..}) = 0.
$$

Thus

$$
E\left[\frac{\sum_{i=1}^{a}(\bar{Y}_{i.} - \bar{Y}_{..})(\bar{Y}_{..} - \mu)}{n\sum_{i=1}^{a}(\bar{Y}_{i.} - \bar{Y}_{..})^2}\right] = 0, \qquad (A2.7)
$$

since the

$$l_i = \frac{(\bar{Y}_{i.} - \bar{Y}_{..})(\bar{Y}_{..} - \mu)}{n \sum_{i=1}^{a}(\bar{Y}_{i.} - \bar{Y}_{..})^2}$$

are for each i identically distributed random variables, from (A2.7),

$$E\left[\frac{\hat{\alpha}'(\bar{Y}_{..} - \mu)}{\hat{\alpha}'(Z'Z)\hat{\alpha}}\right] = E\left[\frac{(\bar{Y}_{i.} - \bar{Y}_{..})(\bar{Y}_{..} - \mu)}{n \sum_{i=1}^{a}(\bar{Y}_{i.} - \bar{Y}_{..})^2}\right] = 0. \tag{A2.8}$$

Now the matrix

$$M = E\left[\frac{\hat{\alpha}\hat{\alpha}'}{\hat{\alpha}'(Z'Z)\hat{\alpha}}\right] \tag{A2.9a}$$

has elements

$$m_{ij} = E\left[\frac{\sum_{i=1}^{a}(\bar{Y}_{i.} - \bar{Y}_{..})(\bar{Y}_{j.} - \bar{Y}_{..})}{n \sum_{i=1}^{a}(\bar{Y}_{i.} - \bar{Y}_{..})^2}\right]. \tag{A2.9b}$$

For the diagonal elements observe

$$r_j = \frac{(\bar{Y}_{i.} - \bar{Y}_{..})^2}{n \sum_{i=1}^{a}(\bar{Y}_{i.} - \bar{Y}_{..})^2}$$

are identically distributed random variables for $1 \le i \le a$. Also

$$\frac{\sum_{i=1}^{a}(\bar{Y}_{i.} - \bar{Y}_{..})^2}{n \sum_{i=1}^{a}(\bar{Y}_{i.} - \bar{Y}_{..})^2} = \frac{1}{n} \tag{A2.10}$$

and, thus,

$$m_{ii} = E\left[\frac{(\bar{Y}_{i.} - \bar{Y}_{..})^2}{n \sum_{i=1}^{a}(\bar{Y}_{i.} - \bar{Y}_{..})^2}\right] = \frac{1}{na}. \tag{A2.11}$$

To obtain the off diagonal elements notice that

$$\sum_{j=1}^{a} m_{ij} = 0. \tag{A2.12}$$

Thus,

$$m_{ii} = -\sum_{j\neq i} m_{ij}. \tag{A2.13}$$

Now m_{ij} are expectations of identically distributed random variables. Thus,

$$m_{ij} = -\frac{1}{(a-1)}\frac{1}{na}. \tag{A2.14}$$

Then from (A2.12) and (A2.14)

$$M = \frac{1}{(a-1)n}T \tag{A2.15}$$

with $T = I - (1/a)J$. Thus, from (A2.6–A2.15),

$$E\left[\frac{\hat{\sigma}^2\hat{\alpha}(\hat{\alpha}-\alpha)'}{\hat{\alpha}'(Z'Z)\hat{\alpha}}\right] = \frac{\sigma^4}{\sigma^2+n\sigma_\alpha^2}\frac{1}{(a-1)n}T. \tag{A2.16}$$

The last term of (A2.1) is a matrix N with elements

$$n_{ij} = E\left[\frac{(\overline{Y}_{i.}-\overline{Y}_{..})(\overline{Y}_{j.}-\overline{Y}_{..})}{n^2[\sum_{i=1}^{a}(\overline{Y}_{i.}-\overline{Y}_{..})^2]^2}\right] = -\frac{1}{(a-1)}n_{ii}. \tag{A2.17}$$

Now

$$\frac{\sum_{i=1}^{a}(\overline{Y}_{i.}-\overline{Y}_{..})^2}{[\sum_{i=1}^{a}(\overline{Y}_{i.}-\overline{Y}_{..})^2]^2} = \frac{1}{\sum_{i=1}^{a}(\overline{Y}_{i.}-\overline{Y}_{..})^2}. \tag{A2.18}$$

Since

$$v = \frac{n \sum_{i=1}^{a} (\overline{Y}_{i.} - \overline{Y}_{..})^2}{n\sigma_a^2 + \sigma^2} \sim \chi^2(a-1),$$

$$(A2.19)$$

$$E\left[\frac{1}{\sum_{i=1}^{a}(\overline{Y}_{i.} - \overline{Y}_{..})^2}\right] = E\left[\frac{n}{(n\sigma_\alpha^2 + \sigma^2)v}\right] = \frac{n}{n\sigma_\alpha^2 + \sigma^2} \frac{1}{(a-3)}$$

$$(A2.20)$$

and

$$n_{ii} = \frac{1}{na} \frac{1}{(n\sigma_\alpha^2 + \sigma^2)} \frac{1}{(a-3)},$$

it follows that

$$p'E\left[\frac{\hat{\alpha}\hat{\alpha}'\hat{\sigma}^4}{[\hat{\alpha}'(Z'Z)\hat{\alpha}]^2}\right]p = \frac{[a(n-1)-2]}{(a-1)a(n-1)(a-3)n} \frac{\sigma^4}{(\sigma^2 + n\sigma_\alpha^2)}T.$$

$$(A2.21)$$

Substituting (A2.16) and (A2.21) into (A2.1),

$$p'E(\hat{\alpha}_{eb} - \alpha)(\hat{\alpha}_{eb} - \alpha)'p$$

$$= \frac{p'Ip}{n}\sigma^2 - \frac{2c(a-1)}{n(a-1)} \frac{\sigma^4}{\sigma^2 + n\sigma_a^2} p'Tp$$

$$+ \frac{[a(n-1)+2](a-1)c^2}{a(a-3)n(n-1)} \frac{\sigma^4}{\sigma^2 + n\sigma_a^2} p'Tp. \quad (A2.22)$$

Now find c so that

$$h(c) = -\frac{2c}{n} + \frac{[a(n-1)+2](a-1)}{a(a-3)n(n-1)}c^2 \qquad (A2.23)$$

is a minimum. Differentiating $h(c)$ and setting it equal to zero yields

$$c = \frac{a(a-3)(n-1)}{(a-1)[a(n-1)+2]}. \qquad (A2.24)$$

The second derivative of $h(c)$ is positive.

Substitute (A2.24) into (A2.22). Then

$$p'E(\hat{\alpha}_{eb} - \alpha)(\hat{\alpha}_{eb} - \alpha)'p$$

$$= \frac{p'Ip}{n} - \frac{a(a-3)(n-1)}{[a(n-1)+2](a-1)}\frac{\sigma^4}{\sigma^2 + n\sigma_a^2}\frac{p'Tp}{n}. \qquad (A2.25)$$

BIBLIOGRAPHY

Albert, A. (1972). *Regression and the Moore-Penrose Pseudoinverse.* Academic Press, New York.

Bannerjee, K.S. and Karr, R.N. (1971). "A Comment on Ridge Regression, Biased Estimation for Non-Orthogonal Problems." *Technometrics,* **13**(4), 895–898.

Berger, J.O. (1985). *Statistical Decision Theory and Bayesian Analysis.* Springer Verlag, New York.

Berger, J.O. and Berliner, L.M. (1986). "Robust Bayes and Empirical Bayes Analysis with ε Contaminated Priors." *The Annals of Statistics,* **14**(2), 461–466.

Bibby, J. and Toutenburg, H. (1978). *Prediction and Improved Estimation in Linear Models.* Wiley, Chichester.

Brown, K.G. (1978). "On Ridge Estimators in Rank Deficient Models." *Communications in Statistics—Theory and Methods,* **A7**(2), 187–192.

Bunke, H. and Bunke, O. (1986). *Statistical Inference in Linear Models.* John Wiley and Sons, New York.

Cassella, G. (1980). "Minimax Ridge Estimation." *The Annals of Statistics,* **8**(5), 1036–1056.

Chawla, J.S. (1988). "A Note on General Ridge Estimator." *Communications in Statistics—Theory and Methods,* **17**(3), 739–744.

Dempster, A.P., Schatzoff, M. and Wermuth, N. (1977). "A Simulation Study of Alternatives to Ordinary Least Squares." *Journal of the American Statistical Association,* **72**, 77–91.

Diderrich, G.T. (1985). "The Kalman Filter from The Perspective of Goldberger-Theil Estimators." *The American Statistician*, **39**(13), 193–198.

Duncan, D.B. and Horn, S.D. (1972). "Linear Dynamic Recursive Estimation from the Viewpoint of Regression Analysis." *Journal of the American Statistical Association*, **38**, 815–821.

Economic Report of the President (1988). Transmitted to the Congress in February 1988 Together with The Annual Report of the Council Of Economic Advisers. United States Government Printing Office, Washington: 1988. (Data taken from tables therein.)

Efron, B. and Morris, C. (1973). "Stein's Estimation Rule and its Competitors." *Journal of the American Statistical Association*, **65**, 117–130.

Farebrother, R.W. (1976). "Further Results on the Mean Square Error of Ridge Regression." *Journal of the Royal Statistical Society*, **B.38**, 248–250.

Farebrother, R.W. (1978). "A Class of Shrinkage Estimators." *Journal of the Royal Statistical Society*, **B.40**, 47–49.

Farebrother, R.W. (1984). "The Restricted Least Squares Estimator and Ridge Regression." *Communications in Statistics– Theory and Methods*, **13**(2), 191–196.

Ferguson, T.S. (1967). *Mathematical Statistics–A Decision Theoretic Approach*. Academic Press, Boston.

Goldstein, M. and Smith, A.F.M. (1974). " Ridge Type Estimator for Regression Analysis." *Journal of the Royal Statistical Society*, **B.36**, 284–291.

Graybill, F.A. (1969). *Introduction to Matricies with Applications in Statistics*. Wadsworth, Belmont.

Graybill, F.A. (1976). *Theory and Application of the Linear Model*. Wadsworth, Boston.

Gruber, M.H.J. (1979). *Empirical Bayes, James–Stein and Ridge Regression Type Estimators for Linear Models*. Unpublished Ph.D. Thesis, University of Rochester.

Gruber, M.H.J. (1985). "A Comparison of Bayes Estimators and Constrained Least Square Estimators." *Communications in Statistics—Theory and Methods*, **14**(2), 479–489.

Gruber, M.H.J. and Rao, P.S.R.S. (1982)."Bayes Estimators for Linear Models with Less than Full Rank." *Communications in Statistics—Theory and Methods*, **11**(1), 59–69.

Haitovsky, T. and Wax, Y. (1980). "Generalized Ridge Regression, Least Squares with Stochastic Prior Information, and Bayesian Estimators." *Applied Mathematics and Computation*, **7**, 125–154.

Harrison, P.J. and Stevens, C.F. (1976). "Bayesian Forecasting (with discussion)." *Journal of the Royal Statistical Society Series*, **B.38**, 205–247.

Harville, D.H. (1976). "Extension of the Gauss-Markov Theorem to Include the Estimation of Random Effects." *The Annals of Statistics*, 4(2), 384–395.

Heffes, H. (1966). "The Effect of Erroneous Models on Kalman Filter Response." *IEEE Transactions of Automatic Control*, **AC-11**, 541–543.

Hoerl, A.E. and Kennard, R.W. (1970a). "Ridge Regression: Biased Estimation for Nonorthogonal Problems." *Technometrics*, **12**, 55–67.

Hoerl, A.E. and Kennard, R.W. (1970b). "Ridge Regression: Applications to Nonorthogonal Problems." *Technometrics*, **12**, 69–82.

Hoerl, R.E., Kennard, R.W., and Baldwin, K.F. (1975). "Ridge Regression: Some Simulations." *Communications in Statistics—Theory and Methods*, 4(2), 105–123.

Humak, K.M.S. (1977). *Statistische Methoden der Modellbildung, Band 1. Statistische Inferenz für Linear Parameter.* Akademie-Verlag, Berlin.

James, W. and Stein, C. (1961). "Estimation with Quadratic Loss." *Proceedings of the Fourth Berkeley Symposium on Mathematics and Statistics.* Berkeley: University of California Press 1, 361–379.

Kagan, A.M., Linnik, Y.V., and Rao, C.R. (1973). *Characterization Problems in Mathematical Statistics.* Wiley, New York.

Kalman, R.E. (1960). "A New Approach to Linear Filtering and Prediction Problems." *Journal of Basic Engineering*, **82**, 34–45.

Kalman, R.E. and Bucy, R.S. (1961). "New Results in Linear Filtering and Prediction." *Journal of Basic Engineering*, **83**, 95–108.

Lauterbach, J. and Stahlecker, P. (1988). "Approximate Minimax Estimation in Linear Regression: A Simulation Study." *Communications in Statistics, Simulations*, **17**(1), 209–227.

Lewis, F.L. (1986). *Optimal Estimation with an Introduction to Stochastic Control Theory.* John Wiley and Sons, New York.

Lindley, D.V. (1965). *Introduction to Probability and Statistics from a Bayesian Viewpoint. Part 2. Inference.* Cambridge at the University Press.

Liski, E.P. (1988). "A Test of the Mean Square Error Criterion for Linear Admissible Estimators." *Communications in Statistics—Theory Methods,* **17**(11), 3743–3756.

Lowerre, J.M. (1974). "On the Mean Square Error of Parameter Estimates for Some Biased Estimators." *Technometrics,* **16**(3), 461–464.

Marquardt, D.W. (1970). "Generalized Inverses, Ridge Regression, Biased Linear Estimation, and Non-Linear Estimation." *Technometrics,* **12**(3), 591–612.

Mayer, L.S. and Willke, T.A. (1973). "On Biased Estimation in Linear Models." *Technometrics,* **15**, 497–508.

Meinhold, R.J. and Singpurwalla, N.D. (1983). "Understanding the Kalman Filter." *The American Statistician,* **37**(2), 123–127.

Milliken, G.A. and Akdeniz, F. (1977). "A Theorem on the Difference of the Generalized Inverses of Two Nonnegative Matrices." *Communications in Statistics—Theory Methods,* **A6**(1), 73–79.

Montgomery, D.C. (1984). *Design and Analysis of Experiments.* Second Edition. John Wiley and Sons. New York.

Myoken, H. and Uchida, Y. (1977). "The Generalized Ridge Estimator and Improved Adjustments for Regression Parameters." *Metrika,* **24**, 113–124.

Neter, J., Wasserman, W. and Kutner, M.H. (1985). *Applied Linear Statistical Models.* Second Edition. Richard D. Irwin, Inc., Homewood, Illinois.

Nishimura, T. (1966). "On the *A Priori* Information in Sequential Estimation Problems." *IEEE Transactions Automatic Control,* **AC-11**, 19–204.

Nomura, M. (1988). "On the Almost Unbiased Ridge Regression Estimator." *Communications in Statistics, Simulation,* **17**(3), 729–743.

Nyquist, H. (1988). "Applications of the Jackknife Procedure in Ridge Regression." *Computational Statistics and Data Analysis,* 177–183.

Peele, L. and Ryan, T.P. (1982). "Minimax Linear Estimators with Application to Ridge Regression." *Technometrics*, **24**(2), 157–159.

Pfeffermann, D. (1984). "On Extensions of the Gauss-Markov Theorem to the Case of Stochastic Regression Coefficients." *Journal of the Royal Statistical Society*, **B.46**(1), 139–148.

Pilz, J. (1986). "Minimax Linear Regression Estimation with Symmetric Parameter Restrictions." *Journal of Statistical Planning and Inference*, **13**, 297–318.

Pliskin, J.L. (1987). "A Ridge Type Estimator and Good Prior Means." *Communications in Statistics—Theory and Methods*, **16**(12), 3429–3437.

Price, M.J. (1982). "Comparisons Among Regression Estimators Under the Generalized MSE Criterion." *Communications in Statistics—Theory and Methods*, **11**(17), 1965–1984.

Pukelsheim, F. (1977). "Equality of Two Blue and Ridge Type Estimators." *Communications in Statistics-Theory and Methods*, **A6**(7), 603–610.

Quenouille, M. (1956). "Notes on Bias in Estimation." *Biometrika*, **43**, 353–360.

Rao, C.R. (1973). *Linear Statistical Inference and its Applications*. Second Editon. Wiley, London.

Rao, C.R. (1975). "Simultaneous Estimation of Parameters in Different Linear Models and Applications to Biometric Problems." *Biometrics*, **31**, 545–554.

Rao, C.R. (1976). "Estimation of Parameters in a Linear Model." *The 1975 Wald Memorial Lectures. Annals of Statistics*, 1023–1037.

Rao, C.R. and Mitra, S.K. (1971). *Generalized Inverses of Matrices and its Applications*. John Wiley and Sons. New York.

Saxena, A. (1984). "Bayesian Estimation of TG Ridge Model." *Statistica Neerlandica*, **38**(4), 256–260.

Schiller, R.J. (1973). "A Distributed Lag Estimator Derived from Smoothness Priors." *Econometrica*, **41**, 775–788.

Schipp, B., Trenkler, G., and Stahlecker, P. (1988). "Minimax Estimation with Additional Linear Restrictions—A Simulation Study." *Communications in Statistics, Simulation*, **17**(2), 393–406.

Searle, S.R. (1971). *Linear Models*. John Wiley and Sons, Inc., New York.

Silvey, S.D. (1969). "Multicollinearity and Imprecise Estimation." *Journal of the Royal Statistical Society*, **B.31**, 539–552.

Singh, B., Chaubey, Y.P. and Dwivedi, T.P. (1986). "An Almost Unbaised Ridge Estimator." *Sankya: The Indian Journal of Statistics*, **B48**, 342–346.

Soong, T.T. (1965). "On *A Priori* Statistics in Minimum Variance Estimator Problems." *Trans. ASME Journal of Basic Engineering*, **87D**, 109–112.

Sorenson, H.W. (1970). "Least Squares Estimation: From Gauss to Kalman." *Kalman Filtering Theory and Application.* Edited by H.W. Sorenson. IEEE Press, New York.

Stahlecker, P., and Lauterbach, J. (1987). "Approximate Minimax Estimation in Linear Regression: Theoretical Results." *Communications in Statistics—Theory and Methods*, **16**(4), 1101–1116.

Stahlecker, P., and Trenkler, G. (1988). "Full and Partial Minimax Estimation in Regression Analysis with Additional Linear Constraints." *Linear Algebra and its Applications*, **111**, 279–292.

Stein, C. (1956). "Inadmissibility of the Usual Estimator for the Mean of a Multivariate Normal Distribution." *Proceedings of the Third Berkeley Symposium on Mathematica, Statistics, and Probability.* Berkeley: University of California Press, 197–206.

Swamy, P.V.A.B., Mehta, J.S., Thurman, S.S., and Iyengar, N.S. (1985). "A Generalized Multicollinearity Index for Regression Analysis." *Sankya: The Indian Journal of Statistics*, **B47**(3), 401–431.

Swindel, B.F. (1976). "Good Ridge Estimators Based on Prior Information." *Communications in Statistics—Theory and Methods*, **A5**(11), 1065–1075.

Teräsvirta, T. (1980). "A Comparison of Mixed and Minimax Estimators of Linear Models." Research Report No. 13. Department of Statistics, University of Helsinki.

Teräsvirta, T. (1981). "A Comparison of Mixed and Minimax Estimators of Linear Models." *Communications in Statistics—Theory and Methods*, **A10**(17), 1765–1778.

Teräsvirta, T. (1986). "Superiority Comparisons of Heterogeneous Linear Estimators." *Communications in Statistics—Theory and Methods*, **15**(4), 1319–1336.

Teräsvirta, T. (1986). "Superiority Comparisons between Mixed Regression Estimators." *Communications in Statistics—Theory and Methods*, **17**(10), 3537–3546.

Theil, H. (1963). "On the Use of Incomplete Prior Information in Regression Analysis." *Journal of the American Statistical Association*, **58**, 401–404.

Theil, H. (1971). *Principles of Econometrics*. Wiley, New York.

Theil, H. and Goldberger, A.S. (1961). "On Pure and Mixed Estimation in Economics." *International Economic Review*, **2**, 65–78.

Theobald, C.M. (1974). "Generalizations of Mean Square Error Applied to Ridge Regression." *Journal of the Royal Statistical Society*, **B.36**, 103–106.

Thurman, S.S., Swamy, P.V.A.B., and Mehta, J.S. (1986). "An Examination of Distributed Lag Model Coefficients." *Communications in Statistics—Theory and Methods*, **15**(6), 1723–1749.

Toutenburg, H. (1982). *Prior Information in Linear Models*. John Wiley and Sons, New York.

Toutenburg, H. (1983). "Insensitivity of Minimax Linear Estimators." *Biometry Journal*, **5**, 501–508.

Toutenburg, H. (1986). *Weighted Mixed Regression with Application to Regressor's Nonresponse. 1. Theoretical Results.* Preprint. Akademie der Wissenschaten der DDR Karl Weierstrass, Institut für Mathematik, Berlin.

Toutenburg, H. (1988). "MSE Comparisons Between Restricted Least Squares, Mixed and Weighted Mixed Estimators with Special Emphasis to Nested Restrictions." *Technical Report*. Akademie der Wissenschaten der DDR Karl Weierstrass, Institut für Mathematik, Berlin.

Trenkler, D. and Trenkler, G. (1981). "Ein Vergleich des Kleinste-Quadrate-Schätzers mit verzerrten Alternativen." In: Fleischman, B., Bloech, J., Fandel, G., Seifert, O., Weber, H. (eds.) *Operations Research Proceedings, 1981*, 218–227.

Trenkler, G. (1983). "Baised Linear Estimators and Estimable Functions." *Scandinavian Journal of Statistics*, **110**, 53–55.

Trenkler, G. (1988). "Some Remarks on a Ridge Type Estimator and Good Prior Means." *Communications in Statistics—Theory and Methods.*, **17**(12), 4251–4256.

Trenkler, G. and Trenkler, D. (1983). "A Note on Superiority Comparisons of Homogeneous Estimators." *Communications Statistics—Theory and Methods*, **12**(7), 799–808.

Vinod, H.D. (1978). "A Survey for Ridge Regression and Related Techniques for Improvements Over Ordinary Least Squares." *The Review of Economics and Statistics*, **60**, 121–131.

Vinod, H.D. and Ullah, A. (1981). *Recent Advances in Regression Methods.* Marcel Dekker, New York.

Wang, S.Q. (1982). "On Biased Estimators in Models with Arbitrary Rank." *Communications in Statistics—Theory and Methods*, **11**(14), 1571–1581.

Zontek, S. (1987). "On Characterization of Linear Admissible Estimators of a Result Due to C.R. Rao." *Journal of Multivariate Analysis*, **23**, 1–12.

AUTHOR INDEX

A

Albert, A., 38

B

Baldwin, K.F., 11, 13
Bannerjee, K.S., 12
Berger, J.O., 15, 228
Berliner, L.M., 15, 228
Brown, K.G., 13, 258
Bucy, R.S., 253
Bunke, H., 15, 16
Bunke, O., 15, 16

C

Chaubey, Y.P., 17
Chawla, J.S., 17

D

Dempster, A.P., 11, 13
Diderrich, G.T., 254

Duncan, D.B., 12, 94

E

Efron, B., 291, 306

F

Farebrother, R.W., 13, 14,
15, 49, 61, 200, 202
Ferguson, T.S., 44

G

Goldberger, A.S., 11, 15
Goldstein, M., 12
Graybill, F.A., 38, 54
Gruber, M.H.J., 14, 15, 95,
104, 120

H

Haitovsky, T., 14
Harrison, P.J., 253
Harville, D.H., 12, 94

SUBJECT INDEX

A

Absolute value, 220
Additional observations, 119
Aerospace tracking, 253, 254, 255
Alternative forms of the
  Bayes estimator, 94-112,
    113, 136, 140, 146, 147,
    160, 168, 236, 250
  mixed estimator, 116-119,
    124, 171-173
  MSE, 165, 170, 185
Analysis of life lengths, 253
Analysis of variance
    (ANOVA), 52, 289,
    290, 295, 309, 320
  table, 297, 310, 312
Applications, 253-326
Approximate Bayes Estimator, 290
Augmented linear model,
    119, 123, 134, 138, 141,
    171, 262, 267

Average MSE
  Bayes estimator, 185, 246, 248
  contraction estimator, 169-170, 320
  empirical Bayes estimator, 290, 304, 319
  Kalman filter, 273-277

B

Basis for column space, 313, 320
Bayes
  estimator, 3, 11, 13, 14,
    19, 38-43, 87-92, 93,
    94, 101, 112, 113, 122,
    125, 147, 165, 167, 168,
    179, 184, 242, 249, 255,
    268, 281, 286, 287, 290,
    302-303, 313-315, 320,
    322
  inference, 253
  methods, 227

339

System dynamics, 281
System equation, 265, 279,
  280

T

Test of hypothesis, 295
Test statistic, 311
Theobald's 1974 result, 46-
  49, 153
Time varying linear model,
  254, 287
Total variance, 6, 11
Treatment mean, 305
Two way classification, 289,
  306-320
Type of prior information,
  151
  augmented linear model
    with constant β, 151
  constant β lying on an
    ellipsoid, 152
  random variable β, 151

U

Unbiased estimator, 10, 156,
  160, 303
Underwater sonar, 253

Updated prior information,
  254, 255, 257, 259, 285,
  287

V

Variance, 87, 167, 168, 180,
  182
  of prior distribution, 159
  average of BE, 159, 165
  of posterior distribution,
    174
Variance components, 312,
  313
Variance   components
    model, 94, 290, 295,
    296, 297
Vector, 119
  space, 28
  subspace, 28, 143, 144

W

Weighted distance, 271
Weighted least square estimator, 69
Weighted mixed regression,
  15, 74, 83

STATISTICAL MODELING AND DECISION SCIENCE

Gerald J. Lieberman and Ingram Olkin, editors

Samuel Eilon, *The Art of Reckoning: Analysis of Performance Criteria*
Enrique Castillo, *Extreme Value Theory in Engineering*
Joseph L. Gastwirth, *Statistical Reasoning in Law and Public Policy: Volume 1, Statistical Concepts and Issues of Fairness; Volume 2, Tort Law, Evidence, and Health*
Takeaki Kariya and Bimal K. Sinha, *Robustness of Statistical Tests*
Marvin H.J. Gruber, *Regression Estimators: A Comparative Study*

MA